Erwin F. Smith

Wilt Disease of Cotton, Watermelon, and Cowpea

Erwin F. Smith

Wilt Disease of Cotton, Watermelon, and Cowpea

ISBN/EAN: 9783337196929

Printed in Europe, USA, Canada, Australia, Japan

Cover: Foto ©berggeist007 / pixelio.de

More available books at **www.hansebooks.com**

BULLETIN No. 17.

U. S. DEPARTMENT OF AGRICULTURE.

DIVISION OF VEGETABLE PHYSIOLOGY AND PATHOLOGY.

B. T. GALLOWAY, Chief.

WILT DISEASE

OF

COTTON, WATERMELON, AND COWPEA

(*Neocosmospora* nov. gen.).

BY

ERWIN F. SMITH,

PATHOLOGIST.

Issued November 22, 1899.

WASHINGTON:
GOVERNMENT PRINTING OFFICE.
1899.

LETTER OF TRANSMITTAL.

U. S. DEPARTMENT OF AGRICULTURE,
DIVISION OF VEGETABLE PHYSIOLOGY AND PATHOLOGY,
Washington, D. C., July 15, 1899.

SIR: I respectfully transmit herewith a report embodying the results of some investigations of a disease which has caused serious loss to the growers of cotton and other Southern crops. The report is technical, and will form a basis for further work looking toward the restriction and prevention of the disease. I respectfully recommend that the report be published as Bulletin No. 17 of this Division.

Respectfully,

B. T. GALLOWAY,
Chief of Division.

Hon. JAMES WILSON,
Secretary of Agriculture.

3

CONTENTS.

ILLUSTRATIONS.

6

WILT DISEASE OF COTTON, WATERMELON, AND COWPEA.

(*Neocosmospora* nov. gen.)

The fungus here described has been under nearly continuous observation for five years. A number of points in its life-history remain to be worked out, but as a year or two must necessarily elapse before the investigation is completed, it is thought best to put the main facts on record at this time, particularly as more material has been accumulated than is usually considered ample for the description of a genus or species.

This material has been accumulating during so long a period that it now amounts to a very considerable mass of correspondence, notes, slides, photographs, drawings, etc., and only a condensed account is here possible. The facts are set forth in as few words as possible, but each one depends on more than a single observation or experiment, and often statements which occupy only a word or two or a line or two are based on scores of observations. Whenever the writer has been in doubt, and that was often, he has repeated the experiment or made additional observations. The long delay in publication is due to the fact that he was disappointed at the result of the cross-inoculations and over certain other failures which are mentioned in their proper place.

DESCRIPTION OF THE FUNGUS.

Ascomycetous stage.—Perithecia on the roots (Pl. I, *1*) more rarely on the parts above ground, superficial, resting on a slight subiculum, sometimes developing in the earth near the roots, or underneath the loose bark of the host plant, or deep in cavities or rifts of the decomposing root or stem, but always free from the tissues of the host and not borne on any distinct stroma,[1] infrequent or numerous, scattered or several to many together; ovate, slightly higher than broad, of quite variable size, ranging from 210 to 400 μ in height by 150 to 328 μ in diameter, mostly 250 to 350 by 200 to 300 μ.[2] Peridium about 20 μ thick, coral red to vermilion red, or in water under a cover glass, of a uniform purplish red, when fully ripe becoming orange vermilion (Ridgway's

---

[1] When cultivated on slices of cooked potato the perithecia are usually developed from a more or less elevated, tough (fleshy), grayish-white stromatic surface of quite different appearance from the ordinary mycelium.

[2] One hundred and eleven measurements.

7

Nomenclature of Colors, 1st ed., Pl. VII, No. 12), owing perhaps to the brown ascospores which can be seen through the translucent walls; collapsing irregularly (more or less) when empty, but not sinking in regularly about the ostiolum after the manner of many Nectrias; surface slightly irregular and a little papillate, especially toward the apex, by reason of the slight projection of single cells, not scaly, pruinose, or hairy; wall distinctly parenchymatic, the irregularly polyhedral cells with rounded angles and in the ventral portion about 8 to 16 μ in diameter (Pl. II, 5); usually no distinct neck until fully ripe, then frequently a short neck (generally 30 to 40 μ long, but shorter or longer, sometimes even as long as 80 μ, but never hairy or fimbriate at the apex);[1] ostiolum closed by special cells (Pl. II, 1) and indistinct until fully ripe, then opening by solution, 30 to 50 μ or more broad, bordered by narrower cells: the neck lined internally by numerous hairs (periphyses), 20 to 30 μ long, which project into the throat leaving only a small opening (Pl. II, 2). Asci 8-spored, numerous (75 to 100 or more), cylindric, stipitate, the somewhat narrowly constricted base of variable length, but usually equal to about one-eighth to one-third the length of the spore-bearing part, the latter 70 to 100 μ long by 11 to 14 μ broad (usually 12 μ broad), with apices rounded and very slightly, if at all, thickened (Pl. I, 2. 3; Pl. V, 2); usually crushing out of the immature perithecium with some part of the hypothecium in one or more adherent masses (Pl. I, 1). becoming so thin-walled as to be scarcely visible, and when fully ripe frequently dissolving, leaving the free spores (Pl. II, 4) to escape through the stoma or through any part of the accidentally ruptured wall as the gelatinous débris in the interior of the perithecium expands by absorption of water, in which case the spores often adhere to the exterior of the perithecium for some time in little irregular brown masses (Pl. I, 1a). In other cases (even on the same culture). the asci retain their form and elasticity and the ripened ascospores are shot out through the ostiolum to the distance of 3 to 8 millimeters. Paraphyses present (Pl. I, 2: Pl. V, 2) and often two or three times the diameter of the asci, each composed of several roundish, oblong, or irregular, thin-walled, loosely connected cells, nearly destitute of protoplasm and readily overlooked (unstained, they are best seen by viewing the sections or crushed out material in water with a very narrow pencil of rays).[2] Ascospores (Pl. I, 2, 3, 5, 6: Pl. V, 2) in one row. when ripe closely filling the ascus, globose to short elliptical, rarely ovate or considerably longer than broad, continuous, rather thick-walled, colorless until ripe, then light brown, in mass mummy brown (Ridgway's Nomenclature, Pl. III, No. 10), with a thick, distinctly wrinkled exospore (more rarely smooth), variable in size. usually 10 to 12 μ in diame-

---

[1] Pure cultures from the same ascospore yielded ripe perithecia with and without beaks.

[2] Similar organs are present in many Nectriaceous fungi which have been described as destitute of paraphyses.

ter, if globose, and 8 to 12 by 11 to 14 μ if ellipsoidal,[1] contents hyaline, one to several guttulate; germinating readily on potato or in nutrient agar (Pl. I, 6), producing a white, immediately much branched, many septate mycelium which by interweaving soon conceals the ascospore and in a short time (cowpea fungus) reproduces perithecia, first, however, and usually within thirty-six hours, producing from the ends of short branches or from terminal hyphae (Pl. I, 7) numerous, colorless, oval to narrowly elliptical conidia which are mostly 8 to 16 by 3 to 6 μ, straight or curved, continuous (or occasionally one-septate after falling off) and indistinguishable from those of the internal fungus (see below). Mycelium on slightly alkaline meat infusion peptone agar at first pure white, then grayish white (stromatic), producing immature but nearly full-grown perithecia in large numbers within a week of the sowing of the ascospore. Mycelium on steamed potato cylinders in test tubes at

---

[1] Altogether 211 ascospores have been measured directly from the plant, and all of the sizes observed are recorded below in microns, together with the plants from which taken and the number of each size. P=Cowpea; W=Watermelon; C=Cotton:

18 x 12 W¹; 17.7 x 10.2 P¹; 16 x 15 P⋅; 16 x 14.5 P¹; 15 x 12 P⋅; 15 x 11.5 W¹; 15 x 10 W¹; 14.7 x 12 P¹; 14 x 14 P²; 14 x 13.3 P¹; 14 x 13 P¹ C¹; 14 x 12 P⁷ W¹ C¹; 14 x 11 P⁴ W¹; 14 x 10 P² W⁴; 14 x 9 P¹ W²; 13.5 x 12 P⁴; 13 x 13 P⁴; 13 x 12 P³ W² C⁷; 13 x 11 P⁴ W¹; 13 x 10 P¹ W¹; 13 x 9 W³; 12.7 x 12 P¹; 12.5 x 10 P⁴; 12 x 12 P³⁵ W¹² C³; 12 x 11 P² W⁵; 12 x 10 P⁵ W⁷ C⁴; 12 x 9.5 C¹; 12 x 9 W¹¹; 12 x 8 P⁴ W⁹; 11.5 x 11 W¹; 11.5 x 10.5 W¹; 11.5 x 10 C¹; 11.5 x 8 W¹; 11 x 11 P² W² C¹; 11 x 10 P² W¹ C⁶; 11 x 9 W⁵ C²; 11 x 8 W⁴; 10.5 x 10 P¹ C¹; 10 x 10 P⁴ W¹⁰ C²⁵; 10 x 9.5 C³; 10 x 9 P¹ W⁵ C²; 10 x 8 W⁷; 9.5 x 9.5 W¹ C²; 9 x 9 W² C¹; 8 x 8 C¹.

On the cotton most of the spores are globose, or nearly so, nearly all have a wrinkled exospore, and the medium and smaller sizes prevail. On the cowpea the majority of the spores are globose and wrinkled, and the larger sizes prevail. On the watermelon a majority of the spores are elliptical, many are smooth and the smaller sizes are the more common, as may be seen from the preceding figures. These variations do not, however, appear to have any specific value, since the spores derived from a single ascospore sorted out and sown on culture media may vary as greatly as any here recorded. This is shown by the following set of figures derived from measurements of ascospores taken from numerous perithecia grown on potato. The pure culture which was used for this purpose was derived from a single wrinkled ascospore taken from a typical perithecium on the cowpea. It will be seen that the general tendency of these spores is to exceed the size of the original spore rather than to fall short of it. They are also larger than the average of those taken directly from the host plants.

Size of 180 ascospores from a luxuriant growth on steamed potato; original spore 12 x 12 μ, the number of each size is given at the right hand above; all with wrinkled exospore except a very few of the small ones: 22 x 12⋅; 20 x 15¹; 18 x 14¹; 18 x 12³; 18 x 11¹; 18 x 10.5¹; 18 x 10⁹; 17 x 12⁹; 17 x 11¹; 17 x 10.5⋅; 17 x 10¹; 16.5 x 12¹; 16 x 15²; 16 x 14¹; 16 x 13¹; 16 x 12⁹; 16 x 11.5¹; 16 x 11³; 16 x 10⋅; 15.3 x 12¹; 15 x 15¹; 15 x 13¹; 15 x 12.5¹; 15 x 12²; 15 x 11²; 14.5 x 13.5¹; 14.5 x 12.5 ; 14.5 x 12⁶; 14.5 x 10⋅; 14.3 x 12.5¹; 14 x 14³; 14 x 13.5¹; 14 x 13⁴; 14 x 12.5¹; 14 x 11.5⋅all from one ascus; 14 x 11¹; 13.5 x 13.5²; 13.5 x 13⁶; 13.5 x 12.5⋅; 13 x 13¹¹; 13 x 12¹⁷; 13 x 10⋅; 12.5 x 12.5⁴; 12.5 x 12⁸; 12.5 x 10.5¹; 12 x 12⁹; 12 x 11.5¹; 12 x 11¹; 12 x 10¹ smooth (7 others from the same ascus were smooth); 11.7 x 11.7¹; 11.5 x 11.5¹ smooth; 10 x 10⁵ smooth. Thirteen of the elongated spores were ovate; the rest were elliptical. A very few spores were flattened on the sides from the pressure of the ascus.

first snow-white, then developing a puffed-up, grayish white, tough stroma which becomes thickly studded with bright coral-red perithecia in about eight days from the sowing of the ascospores. Steamed sterile potato has proved a very suitable medium for the growth of this stage of the fungus, the perithecia becoming ripe and discharging their spores copiously in two to three weeks from the sowing of the ascospores. In a number of instances the cycle from ascospores round to ascospores (ripe and discharged from the perithecium) was as short a time as twelve days and twenty hours. These ascospores were all taken from perithecia found on stems of the cowpea.

Perithecia have also been grown from 150 or more microconidia (Pl. 1,7). They were isolated by the poured-plate method, the material being derived from an agar culture about two weeks old which was made from ascospores. The perithecia developed on five different media, viz, agar, banana, onion, carrot, and potato. On the first three media they appeared sparingly; on the carrot and potato they were abundant and appeared in about two weeks. The ripe ascospores were either shot out of the perithecia against the walls of the test tubes in such numbers as to make distinct brown patches or else were slowly extruded, crowning the ostiolum with irregular brown balls and clumps of spores (Pl. I, 1 a).

So far as known to the writer, this is the second time perithecia have been derived from the conidial fructification of any Hypocreaceous fungus. Brefeld and von Tavel do not record any such case. On the contrary, only three such cases are mentioned by them for the whole group of Ascomycetous fungi.[1] Klebs has since recorded this for *Eurotium repens*, and Hugo Glück, in 1895 (Hedwigia, p. 254), reported it for *Nectria moschata*, which he obtained in about four weeks from pure cultures of the sickle-shaped conidia of *Fusarium aqueductum*, in water and plum decoction to which oak wood and bark had been added. Very likely Brefeld's numerous failures are to be attributed to the fact that he used his "Nährlösung" too exclusively, i. e., did not vary his culture media widely enough, so as to more nearly imitate natural conditions.

On the contrary, the conidial stage of the watermelon fungus (spore taken in July from the interior of a vessel) has been cultivated for five years on a great variety of media, including potato, without showing a

[1] "As many yeast conidia remain under cultivation through endless generations always the same, so ordinarily the spores from conidiophores yield, in pure cultures, always the same conidiophore: those of pycnidia always the same pycnidia. In like manner many oidia, under certain conditions, never produce anything else than oidia. Only in rare cases has it been possible to obtain the ascus fructification from the spores of an accessory fruit form. Of this the Penicillium [crustaceum] spoken of in the second Heft of this work remains the most interesting example. Additional examples are *Endomyces Magnusii* and *Diaporthe controversa*.

"Up to the present time little is known of conditions governing the development of the ascus fructification. Observations in nature frequently show its dependence on a certain duration of development, on the time of year, and on a definite substratum."—Brefeld and von Tavel: Untersuchungen, Heft 10, p. 349.

trace of perithecia, although from time to time special efforts were made
to find a substratum which would lead to the production of perithecia.
This is the strain of fungus which has proved so actively parasitic
in the hands of the writer.

On agar and on potato the perithecia of the cowpea fungus showed a
distinct tendency to be larger than on the host plant, owing probably
to extra good nutrition. The diameter of the largest on nutrient agar
ranged from 320 to 376 μ. The diameter of the largest on potato ranged
from 360 to 416 μ. The perithecia on the potato were exceedingly
numerous, sometimes as many as 5,000 developing on an area of not
more than 15 or 20 square centimeters. They were much less numerous on the slightly alkaline meat infusion peptone agar, and ripened
and developed their bright red color much more slowly. On certain
other media the peridium remained nearly colorless or the perithecia
failed altogether to develop, although the fungus was under the same
conditions except as to the substratum. Numerous cultures on a great
variety of media have shown that the presence or absence of a stroma
is entirely a matter of the substratum.

No attempts have been made to grow the ascospores from the similar
perithecia which have been discovered on the dead stems of cotton and
watermelon. No perithecia ever developed in any of the cultures made
from internal or external conidia taken from the cotton or watermelon.

CONIDIAL FRUITS.

(1) *Microconidia* (Cephalosporium stage).—Colorless, oval to narrowly elliptical, straight or slightly curved, non-septate spores, 4 to 25
by 2 to 6 μ, borne singly one after another (Pl. III, 3) on the ends of
short branches of a mycelium, which fills the water ducts and interior
parts of the *living stem* (melon, cowpea) with a dense growth that is
pure white when seen in mass in the stem or when cultivated out on
alkaline meat infusion peptone agar, or on other alkaline media, even
rice; conidia frequently 1-septate (or rarely 2-septate) after abscission,
in cultures often remaining in a little group around the end of the
stationary conidiophore or becoming scattered when the latter elongates
after having produced a group of spores, as is frequently the case
(Pl. 1, 7, Pl. III, 4), aerial or submerged (*Fusarium vasinfectum* Geo. F.
Atkinson, described from cotton and okra; *Fusarium niveum* Erw. F.
Smith, described from watermelon). None of the large, lunulate, 3 to
5 septate spores described below have ever been observed in the vessels
of any of the host plants or submerged in fluids or solids.

(2) *Macroconidia* (Fusarium stage).—Lunulate; 3 to 5 septate spores,
30 to 50 by 4 to 6 μ (Pl. I, 9; Pl. II, 7; Pl. V, 3), borne on the surface
of *dead stems* in immense numbers on innumerable small, oval or
hemispherical conidia beds which arise from the internal mycelium
and consist of compact, irregularly branched, short conidiophores.
Examined in water the single spores are nearly colorless; in mass they

are at first white, but very soon after the formation of the conidia beds they become colored, the *variable tint* ranging from pink, pale flesh color or pale salmon to deep salmon. When germinated in water or under acid or alkaline agar or in very moist air, producing conidia indistinguishable from those borne by the internal fungus (Pl. II, *10;* Pl. III, *11* in part).[1] Immature spores from the conidia beds are, of course, much shorter than the measurements given above, and for a time are non-septate or only 1-septate. When accidentally knocked off, such spores appear to be still capable of growth and septation. An effort was made to identify this external conidium with some one of the many described forms of Fusarium, but without success. The above measurements are derived from hundreds of spores and are believed to give nearly the limits of variability, but it would be quite possible to make half a dozen species of Fusaria from the material I have had in hand, if only a few measurements were made and a hasty description written out, as is frequently done. Indeed, there is no doubt that the literature of systematic mycology contains much of such rubbish. One has only to study critically a few members of this group and then turn to the descriptions in Saccardo's Sylloge Fungorum to be convinced of it. Not only is the identification of Fusaria from descriptions usually impossible, but very often with the specimens themselves the case is little better, owing to the fact that many Hypocreaceous fungi bear conidia which have been put into the form-genus *Fusarium* and which so closely resemble each other (see the plates in Vol. III of Tulasne's Carpologia, and in Heft 10 of Brefeld's Untersuchungen) that morphology alone is of very little assistance in their identification. In many cases, at least, the only safe way is to make cultures and inoculations if the fungus is a well-marked parasite; or to obtain the less

---

[1] Old cultures of the internal watermelon fungus, when transferred from sterile horse dung to potato broth, produced, along with the small elliptical non-septate conidia, all sizes and shapes up to those which were lunulate, 3 to 4 septate, and 38 μ long, but none 50 μ long could be obtained in this medium, nor were any of them distinctly salmon colored. These cultures were all derived originally from a single non-septate microconidium separated out by the poured-plate method.

When hothouse watermelons were grown in soil infected with pure cultures of the internal melon fungus the macroconidial stage frequently appeared on the surface after the wilting and death of the plants. One of these big, lunulate, 3-septate, external conidia was separated from its fellows and cultivated for a long time, first in sterile horse dung and subsequently in potato broth, in the same manner as the preceding. The mycelium derived from this spore produced great numbers of microconidia, indistinguishable from those borne on the internal mycelium. At least three-fourths of the whole number of spores were of this type, but there were also plenty of typical macroconidia, spores 40 to 50 μ by 4 to 6 μ, lunulate, and 3 to 5 septate. Between these two extremes there were all possible gradations, showing clearly that little dependence can be placed on statements respecting shape and size of spores in the ordinary systematic descriptions of Fusaria. Color is likewise misleading, since the same conidia bed may be white, pinkish, pale salmon, deep salmon, etc., according to its age and the character of the substratum on which it has developed.

13

variable perithecia from pure cultures of the conidia, something we have not yet learned to do with any certainty; or, finally, to discover in nature the perfect fruit form and derive the other forms from it. So far as known to the writer, no Fusaria except those above mentioned have been described from cotton, watermelon, or cowpea; but even if they had, it would in the present stage of our knowledge be next to impossible to establish identity beyond reasonable doubt, since many species of Fusarium are believed to be purely saprophytic and are known to grow on almost any dead substance, and one such form has been discovered by the writer in Washington on a dead stem of cowpea in connection with the perithecia of a Nectria.

(3) *Chlamydospores.*—On the surface of the dead stems of the watermelon and in old cultures of the melon fungus on horse dung, globose, thin-walled, smooth, terminal or intercalary bodies appear, and in mass on the dung are brick red (Pl. III, 12). These are part of the life cycle of the fungus and appear to be chlamydospores comparable with those of *Hypomyces solani* described by Reinke and Berthold.[1] They appear to have an outer and inner wall and germinate readily in water, but have not been studied as critically as the other spore forms. They were found in many test-tube cultures, and have been observed a number of times on the surface of the dead stems associated with the macroconidia. They are usually 10 to 12 μ in diameter, the extreme limits of those measured ranging from 7 to 15 μ.

(4) *Pycnidia(?).*—No pycnidia have been seen either on the host plants or in any of several hundred cultures made on a great variety of media. The cultures began in the summer of 1894 and are still in progress. The examinations on the host plants have included hundreds of dying and dead specimens collected in different years and in various localities from July to October. From these observations the writer believes he is warranted in concluding that there are no pycnidia in the life cycle of this fungus.

EFFECT OF MODIFICATIONS OF THE SUBSTRATUM.

The effect of modifications of the substratum on (1) the production of a stroma, (2) the color of the mycelium, (3) the color of the perithecia, and (4) the development of the perithecia, is described in the following paragraphs:

(1) *The production of a stroma.*—The insignificant subiculum on the host plants and the thick stroma on potato have already been contrasted. The limits of the stroma on potato are beautifully differentiated when thin cross sections are put into chlor-iodid of zinc. The starch-bearing cells of the substratum become blue, while the stroma becomes yellow, and the exact limits of fungus and substratum are easily distinguished. On ordinary nutrient agar the cowpea fungus produces only a feeble stroma. The addition of 1 per cent cane sugar

[1] Die Zersetzung der Kartoffel durch Pilze.

to such agar increased the thickness of the stroma. The stroma was better developed than in the same agar with only 0.2 per cent sugar. In agar with 12 per cent cane sugar there was a great increase of this tissue, which formed a sort of monstrous stromatic mountain the whole length of the long slant. This ridge was 1 centimeter wide and fully one-half centimeter high in its highest parts. Its very irregular, tough, gray-white surface was covered with a thin layer of white aerial hyphæ, and 24 days from sowing was thickly set with colorless perithecia. The latter were, however, also visible at this time on the scanty, non-stromatic mycelium which had climbed up on the walls of the tube above the agar. The formation of perithecia in a similar way was also observed in other culture media. So far, therefore, as this fungus affords any basis for judgment, the stroma is one of the most easily variable parts of a fungus; consequently it appears hazardous to make separation of closely related forms into distinct species solely on the presence or absence of a stroma.

(2) *The color of the mycelium.*—The variability in color according to the nature of the substratum has been a subject of much interest to the writer. In the interior of the host plants the fungus is usually pure white, except in cotton, where the older mycelium is frequently brown. For a considerable time the writer supposed pure white was its only color, but trial of a variety of culture media brought to light some very unexpected and most astonishing facts.

The melon fungus is pure white on slightly alkaline nutrient agar. The fungus remained snow white for over a month on alkaline corn starch steamed in distilled water with asparagin added. In another series it was pure white for 11 days, but had become faint rose color on the fourteenth day and remained so up to the twenty-fifth day.

On boiled rice, with the addition of bicarbonate of soda, the fungus remained pure white for 40 days.

On crushed cowpeas, steamed in distilled water, the fungus made a copious growth from the start, but remained snow white as long as the cultures were under observation (10, 17, and 22 days).

The fungus was pure white in slightly alkaline, peptonized beef broth, and did not become colored on addition of cane sugar or of dextrin.

On acid, neutral, and alkaline nutrient gelatin the fungus made a pure white growth.

In acid potato broth the fungus was white for 11 days, except for certain hyphal strands which had climbed out of the fluid and were attached to the walls of the tubes. The submerged growth and the pellicle were pure white. The following acids were tried with this broth: Malic, oxalic, succinic, tartaric, and citric, 1 c. c. of the $\frac{1}{10}$ normal solution being added in each case to 10 c. c. of the broth. The behavior of the fungus was the same in each case, white on and in the fluid, with brown hyphæ on the walls above it.

On strongly alkaline litmus agar the fungus was white on the sixth day. During the first three days it had made a slow, poor growth, but it now covered the surface with a matted growth from which no hyphæ projected into the air. This medium consisted of 10 c. c. agar, 20 drops of saturated solution of sodium carbonate, and 2 c. c. of Sharp & Dohme's violet litmus solution. On the fourteenth day there was no reddening of the substratum and the surface was covered with a white mycelium, but some of the hyphæ threads which projected out into the air were bluish. At the end of 112 days the agar was nearly dry. It was still blue and was covered by a bluish-white film of fungus. On the glass above the agar many of the filaments of the fungus were now bright blue. This stain most probably came from absorbed litmus rather than from any pigment manufactured by the fungus.

In horse dung sterilized with distilled water the fungus made a white growth (some yellowish hyphæ).

On banana the fungus was white on the start, but in old cultures it was rose, salmon, or purple, and patches of it became bright orange buff.

In 42 days the fungus made a copious growth in the acid juice of Concord grapes, and its color varied from snow white or a very feeble tint (above) to purplish and rose color (below). For some days the fungus was barely able to maintain itself in this very acid medium, and was white, but at the end of 7 days the compact islands of mycelium were roseate. After 42 days the fluid was still strongly acid, and the red color persisted on litmus paper for some years. After 4 months the prevailing tints were white to reddish brown. The very numerous mycelial strands on the walls of the tubes above the medium were a fine red brown (between burnt sienna and tawny).

On Spanish onion cooked in distilled water the fungus made in 3 days a moderate growth, and its color varied from white to lilac (next substratum); in 5 days the fungus was purplish; in 8 days there was a distinct increase of purple and rose color, but there was no color in the substratum under the water. After 21 days the lower part of the onion in both tubes (that part under water) retained its natural yellow color, although a microscopic examination showed it to be grown full of the hyphæ of the fungus. At this time the aerial mycelium was white, rose, purplish, salmon, and brown (a few strands on the walls of the tube). At no time did any crimson color appear.

In 10 c. c. of Dunham's solution with 1 per cent malic acid the fungus filled the fluid and became bright rose color in 20 days, but the acquisition of the color was slow. The litmus reaction was still decidedly acid on the twentieth, thirty-fifth, and forty-fifth days. This experiment was repeated. In 13 days the mycelium was purplish. The addition of 2 per cent cane sugar to this acid Dunham favored vegetative growth, but did not increase the production of color or hasten it. On the fourteenth day in one tube and on the fifteenth in the other the fungus was still pure white. This experiment was repeated.

Sugar did not help on the formation of color, but rather seemed to retard it. The litmus reaction was decidedly acid on the twentieth and twenty-seventh days.

There was marked retardation of growth in Dunham's solution with 2 per cent cane sugar and 1 per cent sodium carbonate. After 20 days the fungus still seemed white, but the fluid was now slightly yellowish. There was less growth than in the acid Dunham. In Dunham's solution containing 2 per cent cane sugar and 0.4 per cent sodium carbonate there was a much better growth on the start than in the same medium with 0.4 per cent of malic acid substituted for the soda. At the end of 13 days the fungus filled the alkaline saccharine fluid and was yellowish brown; the fluid was also browned. The brown color in the mycelium was noticed as early as the eighth day.

The extreme upper part of the gelatin-tube cultures (the substratum) was finally stained rose brown.

In test-tube cultures 1 to 10 of November 27, 1894, on bread steamed in distilled water (5 consecutive days), the fungus grew luxuriantly. On December 18 the bread was nearly or quite covered and hidden by the fungus. In the top of the cultures where it projected into the air the fungus was snow-white. Farther down this pure white color shaded into yellows, purples, flesh tints, and reds, the predominating color being an irregularly distributed bright crimson. All of the tubes showed this crimson color, some more than others; several were very striking. There was less purple and more crimson than in the same fungus when grown on potato.

In 6 days, on pearl tapioca steamed in distilled water, the fungus was purple.

In 3 days, on crushed wheat steamed in distilled water, there was a copious growth and a bright purple seam where the fungus rested on the substratum. In 6 days from the inoculation the purple color had extended and a brilliant crimson was visible in the hyphæ bordering on that part of the substratum first attacked. The fungus resting on the wheat was purple, while the upper parts of the same mycelium were pure white.

In 3 days, on hominy steamed in distilled water, the fungus made a copious growth and developed an abundance of color. This was purple in the greater portion, but crimson in the oldest part. On the sixth day the purple color involved the upper half of the substratum in one tube and nearly the whole of it in the other tube.

The rice cultures proved particularly interesting. On the third day, in tubes of rice steamed in distilled water, there was a copious development of mycelium, which was snow-white in the air and purplish and crimson on the substratum. In 5 days the bulk of the culture was purplish to carmine. Even the aeriel mycelium was tinged. A little of the mycelium in the bottom of the tube and some on the walls of the tube above the rice was still colorless. After 20 days this crimson rice was stirred up in distilled water which became pink even when 20 to 30 c. c. was used with the contents of but one test-tube culture. This fluid was strongly acid to litmus. On filtering it through tissue paper it was colorless. On boiling it gave off a pleasant aromatic odor and litmus paper held in the steam was reddened. Most of the acidity passed away on boiling. The color of the unfiltered fluid changed from pink to purple on boiling in a sterile test tube. There was no change of color with 1 per cent chromic acid.

Six rice cultures were now instituted in test tubes as follows, each, with one exception, holding about the same quantity of the cooked rice: (1) Rice with 2 drops of saturated solution of sodium carbonate; (2) rice with 3 drops of the soda solution; (3) rice with 5 drops of one-half per cent solution of hydrochloric acid; (4) rice with 10 drops of this acid water; (5) rice made blue with 2 c. c. of violet litmus solution; (6) rice made blue by 2 c. c. of violet litmus solution and 2 drops of saturated solution of sodium carbonate. These tubes were all inoculated in the same way, from one of the crimson rice tubes just described, and were under the same cultural conditions. They will be described by number. In 2 days most of the mycelium in 1 was snow-white, but there was a slight purpling around two rice grains; in 2, which differed from the others in having only about two-thirds as much rice and was, therefore, nearly twice as alkaline as 1, the fungus was growing rather slower and was pure white throughout; in 3 there was a decided purpling of most of the fungus and of the rice grains which were touched by it; 4 was like 3; in 5, that part of the fungus projecting into the air was pure white, but the blue rice grains in the vicinity of the growing fungus had become bright purple; in 6 there was also some purpling of the rice grains attacked, but it was more restricted, and, as in 5, the bulk of the rice was bright blue. On the 6th day the condition was as follows: 1, fungus has made a good growth and some of it is pale purple to light rose, but there is not one-tenth as much color as in the tubes which received the acid; 2, the fungus has grown much within the last few days, but there has been no corresponding increase of the color, which is now a faint, scarcely noticeable rose or purple, i. e., there is not one-hundredth part as much color as in the acid tubes; 3 and 4, very brilliant and beautiful, the prevailing color, crimson, shading into rose and purple, but where the fungus projects above the culture into the air it is pure white, or shows only the faintest tinge of pale rose; 5, the upper one-half of the tube is filled with a copious growth of the fungus, and most of the blue color has disappeared, being changed below into violet and above into rose color, and brilliant purples and crimsons, but that part of the fungus which grows into the air above the culture is white to faint rose; 6, like 5, except that the lower one-half of the rice is still a pure blue. On the 9th day tubes 1 and 2 offered a striking contrast to 3 and 4. In tube 1 the fungus was becoming quite roseate over large areas, but none was crimson or purple: in tube 2 (which received twice as much alkali) the fungus had grown through all parts of the rice, and varied from snow white to a faint roseate color, which was scarcely noticeable. In 3 and 4 the starch grains were covered and hidden by the fungus, which was snowy only where it projected into the air above the culture. There were no purple hues, but the body of the culture varied from rose to crimson, the latter color predominating. There was now most crimson in the tube which received the most acid, but possibly this was accidental. In 5, all of the blue had disappeared, but the rice grains at the extreme bottom of the tube still showed a very little red purple. All the rest of the substratum was so closely invested by the fungus that the rice grains were hidden. The color of the fungus varied from rose color to the most vivid crimson, the latter color predominating, the only white part of the fungus being that which projected into the air at the top of the culture. In 6, which was more alkaline, a half cubic centimeter of rice in the bottom was still violet color, none was blue, and in other respects the culture was like 5. After 40 days 1 and 2 still showed the restraining effect of the soda. In 1, a little of the mycelium was now crimson, but most of it was only pale rose color; in 2, there was much less color than in 1, most of the color being pale rose, but with a little violet and a trace of crimson; the rice had shrunken from one side of this tube and the cavity was filled with white mycelium. In 3 and 4, crimson was still the prevailing color; both were shrunken away from the wall on one side, and in one tube white mycelium was beginning to fill the cavity. In 6, the lower one-third of the rice was now brilliant crimson, and the upper two-thirds, which had lost much water, was changed to dark purple and partly overgrown with white mycelium. The dark purple color and the crimson could be seen distinctly in the hyphal strands as well as in the rice grains themselves. In 6, at the end of 58 days, the entire mass was shades of violet, no crimson color remained, and the secondary superficial mycelium was white.

The above experiment was repeated as follows with five test tubes of rice boiled in distilled water: (1) Rice, with 3 drops of saturated solution of sodium carbonate; (2) rice, with 6 drops of the soda solution; (3) rice; (4) rice, with 5 drops of one-half per cent hydrochloric acid; (5) rice, with 10 drops of one-half per cent hydrochloric acid. All were inoculated at the same time in the same way and from the same culture. In 6 days there was a large amount of color in tubes 3, 4, and 5—rose, purple, and crimson. The color was as well developed in the pure rice as in the tubes which received the acid. Tubes 1 and 2 behaved much as in the preceding series. In 1, at the end of 14 days, the entire rice was invaded, and some of the mycelium was roseate to faint purplish, but there was not one-hundredth part as much color as in the acid tubes. In 2, the whole substratum was invaded, and the mycelium was white to faint roseate, there being about one-fourth as much color as in 1. On the twenty-fifth day, in 1, the color was much deeper than on the fourteenth, but it was still mostly white or roseate, there being very little crimson. In 2, the fungus was white to rose color, with occasional hyphal strands becoming crimson. The color had increased, but it was only about one-half as much as in 1. In 3, the body of the culture was a beautiful crimson mottled with purple. The color was lodged both in the substratum and in the hyphal strands. Where the rice had shrunken away from the wall a net work of white mycelium was growing out of the crimson mycelium and covering it. In the upper part of the tube, where the fungus projected into the air, it was also white, or rather, so nearly so that it appeared tinted only when taken out and massed on white paper. The conditions in 4 and 5 were like those in 3, except that there was no purple color. In both tubes the rice was shrunken away from the wall and white mycelium was filling the cavities. This contrasted curiously with the crimson hyphæ and rice grains from which it was growing.

A tube of litmus rice made much more strongly alkaline than the two previously mentioned was inoculated October 28, 1895. On November 14 the upper one-third of the rice (that best aerated) was thoroughly invaded by the fungus, the deeper parts being free, or nearly so. The fungus was snowy white. The rice grains invested by the fungus were changed from bright blue to a pale blue, but there was no purple or red color. This culture consisted of 15 c. c. cooked rice, 2 c. c. of violet litmus solution, and three-fourths cubic centimeter of a saturated solution of sodium carbonate. On December 5 the rice had lost all of its bright blue color, but there was no violet or crimson. The entire cylinder of rice was more or less overgrown and interwoven with the fungus, which was white; the rice grains in the upper part of the tube were greenish gray, and in the lower part a pale purplish rose.

In cutting through old dense conglomerated rice cylinders overgrown with the crimson fungus, the periphery was generally crimson, while the center of the cylinder was mottled, the exterior of the rice grains being red and their interior white. This seems to be another indication that the pigment requires an abundance of free oxygen for its development.

Three large Erlenmeyer flasks of rice were prepared as follows: (1) One-half pound of rice sterilized in several hundred cubic centimeters of distilled water; (2) one-half pound of rice boiled in several hundred cubic centimeters of distilled water, then knocked out of the flask and intimately mixed with one-fourth pound of c. p. calcium carbonate, and subsequently resterilized with more water; (3) one-half pound of rice boiled in several hundred cubic centimeters of distilled water, then 100 cc. of a saturated solution of sodium carbonate added, and the sterilization completed. These flasks were inoculated at the same time, in the same way, and from the same culture. In 48 hours the condition was as follows:

In 1, the fungus had begun to make a vigorous growth from four different centers of infection; color, purplish for the most part, but some

crimson. Two rice grains which projected up above the general surface, and which were invested by the fungus, were wholly deep purple
verging to crimson, while the hyphæ projecting from them were colorless (white).

In 2, nearly as much growth as in 1, but much less color, the hyphæ
being white above and purplish where they rested on the colored substratum, but not crimson.

In 3, fungus pure white, but only a trace of growth; i. e., not one
two-hundredth as much as in 1 or 2.

At the end of 13 days the condition was as follows: In 1 the surface
was nearly covered with beautiful masses of crimson mycelium; the
upper half of the rice was also invaded by the fungus and was mottled—
white, purple, and crimson; 2, too much water in this flask, and fungus
growth conspicuous only on the surface; no crimson stain, but some purple hyphæ; nearly all of the fungus was snow white; in 3, fungus snow
white; there were now five small tufts of fungus on the surface of the rice,
but the latter was too alkaline for a rapid growth. In 21 days the condition of No. 3 had much improved, the fungus being visible on a hundred
or more of the surface grains of rice as a thin white hyphal layer. On
the thirty-seventh day the fungus had increased noticeably in growth,
and all of it was pure white, and the rice grains were also free from
color. That the alkali did not simply mask the color was shown by the
fact that when some of the fungus-infested rice grains from this flask
were put into dilute hydrochloric acid there was no change of color. In
flask 1 the fungus had not reached the bottom of the rice at the end of
21 days, but there were traces of it nearly to the bottom. The upper
half of the culture (a large mass, it will be remembered) now varied from
rose color to crimson, both the rice and the fungus, and was very showy.
After about 40 days, during which the bright color increased in some
parts and faded to purple in others, the top part of flask 1 was extracted
with hot alcohol. The alcohol became brown-red and a bluish-violet
residue remained. This brown-red alcohol became opalescent on the
addition of water, but the precipitated color passed readily through a
filter paper. The alcoholic extract faded slowly on exposure to light.

On another occasion some of the bright-red fungus was extracted
with alcohol. In quantity the alcoholic extract was dragon's-blood
red, and the residuum was blue-purple. The red alcoholic extract was
rendered colorless by a few drops of strong caustic-potash liquor or
caustic-soda liquor, but was not destroyed by liquor-ammonia fort., even
when used in large quantities. The addition of a saturated solution
of sodium carbonate did not destroy the color. It would seem, therefore, that two colors may be present—a blue and a red—the latter soluble in alcohol and easily destroyed by light and by caustic potassa and
soda, the former insoluble in alcohol, more resistant to light and
unaffected by alkalies.

The red fungus from a rice culture was tested with various reagents

and the results are here given for comparison with those obtained with
the red color of the perithecia, which are given below: *Strong ammonia*, exposure to the vapor, even for an instant, changes the red to
purple; fragment still purple after 42 hours in the liquor; fluid itself
slightly tinged. *Caustic-soda liquor*, red changed instantly to very
dark purple, almost black; purple after 42 hours; fluid slightly tinged.
Caustic-potash liquor (strong), red changed immediately to purple; after
42 hours still purple; lower one-third of the liquid reddish—i. e., much
more decidedly tinged than any other alkaline fluid. *Sodium carbonate*
(saturated solution), red changed at once to purple; purple after 42
hours, and the fluid no longer quite colorless. *Ammonium carbonate*
(saturated solution), red changed at once to purple—it changes even
in the vapor; purple after 42 hours; fluid as in the preceding—i. e., a
very slight departure from colorless. *Chloral hydrat* (saturated solution), no immediate change unless to become brighter red; soon soluble;
after 42 hours fluid still roseate; fungus dull red. *Sulphuric acid* (1
per cent), no change in 42 hours; fluid colorless, fungus red. The same
result was obtained with 10 per cent sulphuric acid. *Nitric acid* (1 per
cent), no immediate change, and none after 42 hours; fungus still red
and fluid colorless. *Hydrochloric acid* (1 per cent), no change in 42
hours; fungus red, fluid colorless. *Acetic acid* (1 per cent), no change
in 42 hours; fungus still red and fluid colorless. *Wood alcohol*, distinctly soluble; alcohol around the red fragments of fungus colored
within a few minutes; after 42 hours fungus purple, fluid pale wine-red.
Ethyl alcohol (common, 95 per cent), color soluble; the alcohol around
the fungus became colored in a few minutes; after 42 hours fungus
purple, fluid a pale wine-red. *Sulphuric ether*, no immediate action, but
after an hour or two the ether was tinged; after 42 hours fungus purplish, fluid pale brownish-red. *Chloroform* (purified), no immediate
change, and none after 1½ hours; after 42 hours fungus purplish, fluid
pinkish-red. *Benzine* (purified), no change in 43 hours; fungus still
red and fluid colorless. *Benzole*, no change in 1½ hours; after 43 hours
fungus red to purple-red, and the fluid with a slight pinkish tinge.
Carbon bisulphide, no change in 43 hours; fungus still red and fluid
colorless.

One test was made to determine whether the fungus would produce
its colors on uncooked rice in water. In 2 days a distinct purple color
appeared on the rice grains infested by the fungus.

All of the above experiments relative to color production were made
with pure cultures of the melon fungus.

Allusion has already been made in several places to the growth of
white mycelium from colored, and vice versa. It happens very frequently that on mycelium of the same size and apparently of the same
age one branch will be colorless (white) and another deeply stained
(purple, red, blue, brown, yellow). The stained hyphæ were usually
quite granular, but unstained hyphæ were also sometimes observed

with granular contents. Went observed the same thing in Monascus (Ann. des Sci. nat. Bot. 1895, No. 1), and the writer has also seen it in that fungus.

The cowpea fungus also stains various starchy substrata, but the writer was never able to get as bright colors as with the melon fungus.

The red color of the perithecia is insoluble or nearly so in ether, chloroform, benzine, benzole, or carbon bisulphide. The color is very slowly soluble in ethyl and methyl alcohol, but decidedly more so in the wood alcohol than in the 95 per cent ethyl. That is, after 48 hours there was no change in the ethyl alcohol, and even after several weeks most of the perithecia were still bright red, although a few had faded; after 48 hours in the methyl alcohol the red perithecia (from a potato culture) were a little paler, and after several weeks a few only were pale red, most of them being pure white. In a saturated solution of chloral hydrat the color changed at once slightly, was dull orange-red in 15 or 20 minutes, yellowish-red in 2 hours, and dull yellow in 48 hours. In chlor-iodid of zinc the color soon changed to dark purplish red and then to brown. The mycelium and interior of the perithecia did not blue, but changed from white to yellow. In mineral acids (nitric, sulphuric, hydrochloric, chromic), strong or weak (weakest solution 1 per cent), the perithecia changed instantly from red to yellow and remained yellow. The same change took place in acetic acid, but required several minutes when a 1 per cent solution was used. In a saturated solution of ammonium carbonate the perithecia were purple at the end of 48 hours, but there was no visible change during the first 2 hours. In strong ammonia water they changed immediately to pansy-purple, and did not at once lose color, but were paler at the end of 48 hours.

A purplish stain has frequently been observed in the parenchyma of old dead stems of cotton, melon, and cowpea attacked and killed by this fungus, and this stain when subjected to certain solvents behaves in the same way as the red pigment of the perithecia.

The formation of purple or other bright colors in the substratum, and occasionally in the mycelium itself, appears to be common to many Fusaria. Schacht observed it many years ago (Prings. Jahrb., Bd. III, pp. 446–447). In potatoes attacked by wet rot he found the cavities lined by *Oidium violaceum* Harting. This lining was dark violet to blue black. This color he states to be due to the contents of the mycelium and to the effect of the latter on the substratum. On addition of sulphuric acid the mycelium and spores became rose red. This change, he thought, denoted conversion of starch into sugar, for it took place even inside of starch grains, and the color is similar to that which may be obtained with sugar, albuminoids (Eiweissstoff), and sulphuric acid. In dry starch attacked by the fungus and kept since 1855 he could, however, get no reduction of the copper on warming in Fehling's solution. *Oidium violaceum* appears to correspond to the chlamydospore stage of Neocosmospora. Schacht states that it is only one stage of *Fusisporium solani*, he having found both spore forms attached to the same mycelium.

K. Klein (Beitrag z. Kent. d. rothen Malzschimels, Mitt. d. österr. Vers. Stat. f. Brauerei u. Mälzerei, Wien, V, 1892) has also found a Fusarium causing bright color reactions on starchy media, but I have not seen his paper.

My conclusions relative to the formation of color by the melon fungus are as follows:

(*a*) On neutral or acid media in the presence of free oxygen and of starchy foods—e. g., potato, bread, rice, tapioca, wheat, hominy, cucumber agar, etc.—this fungus develops in the substratum a series of the most brilliant colors (Pl. I, *13*), which are then absorbed by the hyphæ. These hues include many shades of pink, red, purple, and violet, and in some of the substrata—e. g., bread or boiled rice—are particularly brilliant, changing gradually from shades of purple and rose color into the deepest crimson (rose carthamine). This color is much brighter and purer than any I have been able to obtain with Went's *Monascus purpureus*. During the development of this pigment the substratum becomes intensely acid (mostly CO_2, but some lactic acid according to Mr. K. P. McElroy). If, however, alkaline substances (caustic lime, carbonate of soda, etc.) be added to the substratum in advance, so as to neutralize the acid or acids as fast as formed, no color is developed, the fungus remaining snow white, as in the vessels of the melon plant. If less alkali be added, the colors appear gradually after a time, which is longer or shorter according to the amount added.

(*b*) The yellow and brown colors are formed in the presence of an alkali, but apparently not unless sugar is also present. According to the writer's view, the brown stain of the lignified walls of the vessels is due beyond question to the presence of this fungus. Since the lignified walls are much more apt to be stained than the pure cellulose walls, it would seem as though the presence in the lignified wall of coniferin or of some related substance may have something to do with the production of the brown stain. The vascular bundles in melon, cotton, and cowpea contain a distinctly alkaline fluid and the fungus is able to dissolve its way through cellulose walls. If it should also be able to split up coniferin, or some similar glucoside, held in the walls of the vessels, with the liberation of a sugar, then a brown stain might perhaps be formed just as readily inside of the plant as in the alkaline peptone water to which cane sugar was added.

(3) *The color of the perithecia.*—No white-walled or colorless perithecia have ever been observed on any of the host plants. All have been bright red. They have also been red on a variety of artificial media, particularly those on which they have grown best—e. g., steamed potato, crushed wheat, malic acid agar. Red must therefore be assumed to be their natural color.

The color of the peridium is, however, much more easily modified than that of the ascospores. The latter have always remained brown, no matter on what medium grown: the former have not always developed a red color. In a few instances the writer has succeeded in growing perithecia of the cowpea fungus to maturity, and the shedding of ascospores, without the development of any red color in the peridial wall. In a number of other cases the red color has been very slow to appear. It is thus established beyond doubt that the color of the peridium of a species may be entirely a matter of the particular substratum on which it happens to be growing. The principal differences observed in my cultures are given below:

(1) On steamed potato the fungus fruits copiously, the color of the peridium comes quickly (sixth to ninth day), and is first roseate, then bright coral red. The ascospores are mature in two to three weeks. By the sixth day the substratum has changed from white to dark lead gray. In one instance only (out of a great many) on the upper (dry) end of a potato culture 10 days old many perithecia were still nearly colorless.

(2) On pearl tapioca steamed with distilled water the fungus grows slowly[1] and fruits sparingly; the color of the peridium comes slowly, being first pale red and finally *black*. At the end of $2\frac{1}{3}$ months most of them were destitute of necks, but a few of the black perithecia in one tube had distinct necks. Some of the blackest were removed and examined microscopically. They contained no ascospores or asci. The peridium was brown, but normal in structure. The interior of the perithecia was very full of oil. Some substance necessary to the formation of ascospores appeared to be wanting in this substratum.

(3) On yellow banana steamed with distilled water the fungus grew slowly and fruited only after several weeks, but then copiously. At the end of 5 days there was not one-fiftieth as much vegetative growth as in corresponding tubes of potato. The perithecia were pale yellowish red. None were bright red, and some were nearly colorless.

These remarks apply to two cultures made November 13, from the same culture, and examined December 16. In other cultures the perithecia were ochraceous buff at the end of 20 days. On January 23 the two tubes of November 13 were reexamined and found to be quite unlike. In one the perithecia were now coral-red, had decided necks, and were discharging ascospores—i. e., protruding them through the ostiolum in brown masses. In the other tube, the shape and the color of the perithecia were both aberrant. This culture was covered all over with perithecia, but these were globose and destitute of necks: a few were pale coral-red, but most were yellowish brown to dark brown. In the darker ones the structure of the peridium could not be made out without crushing the perithecium. There were no ascospores in this culture, not

[1] The amount of growth at the end of 8 days was not one one-thousandth as much as on potato.

even in the dark-brown perithecia. One of the darkest perithecia was crushed in water. It contained no asci, but every cell was filled with oily looking globules which blackened with osmic acid. Here we have the substratum causing change in color, change in form, retardation of the production of asci, and the storage of large quantities of reserve material. Most of the old banana cultures resembled this one rather than the preceding. The dark brown or black color was lodged in the walls of the peridium. The oily granules gave a milky appearance to the crushed contents. In some of these tubes certain microconidia were also observed to have changed into round spores. In all, 17 banana cultures were under observation, and for a long time. The one first mentioned seems to have possessed a trace of some substance wanting in the others, and which finally enabled the fungus to fruit normally.

(4) On Spanish onion steamed in much distilled water the fungus fruited slowly, but finally quite freely, the perithecia being pale red to bright coral-red.

(5) On commercial (alkaline) cornstarch steamed in distilled water the fungus grew sparingly and fruited sparingly. At first the perithecia were pale red to coral-red, but after a month they were dark—i. e., almost as black as on the tapioca. Even after 2½ months there were very few perithecia. Some were pale red, but most of them were dark brown, especially the older ones. There were no asci in these dark-brown perithecia, but great quantities of oil globules. The mycelium was also modified. The latter was extremely variable in thickness (1 to 13 μ), irregular, much inclined to form swollen places and globose ends, or moniliform chains or clumps of big rounded cells. It was also densely packed with refringent globules. In two of the black perithecia which were crushed in water the peridial wall was brown under the microscope, but normal in structure.

(6) On crushed wheat steamed in distilled water the fungus fruited quickly and copiously, but the perithecia were duller red than on potato. In 8 days the surface was densely packed with dull-red perithecia. At the end of a month some of them which were discharging ascospores were nearly colorless, while others, also discharging ascospores, had a tinge of purple (not red). Other large clumps, which appeared to be full grown, were entirely without color—i. e., a dirty white. Most, however, were colored on this substratum. At the end of 2½ months there was a copious protrusion of ascospores. Nearly all of these were globose, brown, with a thick wrinkled epispore, and were 11 to 13 μ in diameter. The interior of these spores was guttulate. A very few of the ripe spores were: (a) smooth, (b) broadly elliptical, (c) ovate. Several hundred were examined in vain for a septum. At this time the walls of many of the perithecia, protruding spores, were nearly color-less—very pale red or yellowish white.

(7) On hominy steamed in distilled water the fungus fruited slowly

and in moderate abundance. In 16 days the entire substratum was dark purplish red; the perithecia, yellowish red. At the end of a month the perithecia were of good size, but many of the ripe ones (discharging brown ascospores) were colorless or pale yellow and the rest were yellowish red. None were coral-red or bright red of any shade, and the majority were pale yellowish white—i. e., different enough to be considered another species, as species are often made. After 2½ months one of these cultures was still very interesting. The cylinder of hominy, which was now about 3 centimeters high. had shrunk away from the walls of the tube and its surface bore about a thousand normal-shaped perithecia. A majority of these had protruded ascospores which adhered in irregular brown masses to the top of the neck around and over the ostiolum. The striking fact was that *not one of these ripening and ripe perithecia was red; all were pale or dirty yellowish white,* including those crowned by the ascospores. Another culture of the same date and kind was exactly like this one, except that it contained about twice as many perithecia, not one of which was red; all were dirty or pale yellowish white. Many were shedding ascospores, but not so large a number as in the preceding tube. Nothing could be plainer than that the fungus was unable to extract a red color from this substratum; still it was able to ripen its ascospores normally.

(8) The color of the peridium was very slow to appear in perithecia on alkaline agars, and on the same with the addition of cane sugar. After 75 to 85 days, only a few were bright red, while many were colorless or only faint roseate or yellowish.

(9) On rice steamed with distilled water (5 drops of very dilute hydrochloric acid added to each test tube) the substratum was dark purplish red at the end of 16 days, except for a few white grains which projected above the rest. The perithecia were numerous and full size, but ranged from colorless to yellowish red. None were vermilion or coral-red. On the surface of a thick rice gruel no perithecia had formed at the end of 24 days. The body of the gruel, which was in a small Erlenmeyer flask, had become dirty white; its surface was mottled brown and pale rose color. At the end of 38 days the body of the rice was light ecru drab, with possibly a trifle of rose color in it. The surface was thickly spotted with brown flecks, the largest of which bore numerous reddish-brown perithecia. In a few places white hyphæ were pushing out into the air in small numbers. At the end of 65 days the fungus in this flask had formed a nearly continuous dark-brown surface mat, which bore numerous perithecia, some of which had shed ascospores. The color of the peridium varied from pale yellowish brown to reddish brown, and even to dark brown. The body of the rice was rather feebly stained—vinaceous cinnamon mixed with a little vinaceous pink.

(1) *The development of the perithecia.*—By the use of strongly alkaline media the writer found it possible to entirely prevent the formation of

perithecia in the cowpea fungus, although the vegetative growth and the formation of the microconidia continued. This is the more remarkable because on all ordinary media the fungus showed the strongest tendency to form perithecia, even when it was unable to ripen its ascospores. The results obtained are as follows:

(1) In 92 days, on strongly alkaline beef-broth peptone agar, copiously inoculated with fresh ascospores, there was no trace of perithecia. This media consisted of two test tubes of standard nutrient agar (stock 82a). One of them, which contained 10 c. c. of agar, received 10 drops of a saturated solution of sodium carbonate (temperature 25° C.), and the other, which contained 8 c. c. of agar, received 4 drops of the same soda solution. Each was resteamed on 3 consecutive days. Boiling with this alkali caused a heavy precipitate in each tube and a dark color in the agar which received the most soda. Growth appeared first in the tube which received the least alkali. Neither showed any for the first 3 days. On the sixth day the slant surface of each was spotted all over by the fungus. Growth, however, in each was sparing, considering the number of ascospores used and the time which had elapsed. In the tube which received the least alkali the growth was whiter, and there was more fungus in the air than in the other tube, where the growth was mostly in the substratum. At the end of 10 days, in the tube which received the most alkali, the surface was gray-white, with scattered white flecks. The growth of the fungus was still mostly in the substratum and was not copious enough to hide the surface of the agar. The tube which received one-half as much alkali presented the same general appearance. At the end of 26 days both tubes were almost exactly alike; the entire surface of each was covered with a grayish stroma sparingly flecked with white. In other words, the growth of the fungus, while not so prompt as on neutral or feebly alkaline agar, was not prevented by the strong alkali. At the end of 51 days in both tubes the mycelium showed a marked tendency to break up into iso-diametric, or nearly iso-diametric cells, constricted at the septa and rounded in the middle. In both tubes microconidia were now abundant, and it was thought that possibly perithecia might develop later on, as there appeared to be tiny hyphæ-complexes in some of the mycelium, which had crept up out of the alkali bed and was now growing on the walls of the tube.

These cultures were started November 30, and up to January 24 they were exactly alike; that is, the surface was covered all over by the fungous stroma, which was grayish white and bore numerous microconidia, but no perithecia. On this date 2 c. c. of a sterile 2-per-cent solution of malic acid was pipetted into the tube which had received the least alkali. *In 10 days 1,500 unripe but well-developed and bright coral-red perithecia made their appearance.* The whole surface of the corresponding tube (alkaline agar) was covered with a grayish-white sheet of fungus, mostly in the agar (stroma), and it still bore an abundance of microconidia, *but there was not a single perithecium.* Up to the ninety-second day, when the experiment was lost, not a single perithecium had formed on this alkaline agar. No contrast could be more striking, except possibly the following:

(2) On the best medium known to the writer for the abundant and rapid production of perithecia, viz, potato steamed in distilled water, the development of perithecia was altogether prevented by the addition of carbonate of soda. A copious stroma was formed, and there was an abundant development of microconidia, *but in 76 days there was no appearance of perithecia.* These cultures were on potato cylinders steamed in test tubes with 2 c. c. of water and 1 c. c. of a saturated solution of sodium carbonate. A second set of tubes was prepared in the same way, but received less alkali (about one-half as much). For comparison, cylinders cut from the same tuber were steamed in a strong malic-acid water. The acidified cylinders were white; those steamed with the alkali were uniformly brown. The three sets were inoculated at the same time (December 16) copiously with ascospores from the same culture. On the fourth day there was a much better growth in the acid tubes; growth in the alkaline tubes was sparing and only to be detected with a hand lens.

On the tenth day the fungus in the alkaline tubes had made a pure white but very restricted growth, varying in area from one-half to 1½ qcm. In the tubes which had received the least alkali there was a stroma and an abundance of microconidia, but in none of the four was there any trace of perithecia. On the malic-acid substrata the fungus had now entirely covered and hidden from sight the whole of the large potato cylinder with a wrinkled stroma. This bore in each of the four tubes several thousand perithecia, which were already rose color or bright coral red. The acid appeared to have favored rather than hindered their formation. On the seventeenth day the entire surface in each of the acid tubes was so thickly studded with bright coral-red perithecia that there seemed to be no room for any more. These tubes were now no longer acid, although strongly so when inoculated. Each of the four cylinders was tested. There was no trace of acid reaction in two, and in the other two it was extremely faint and limited to the bottom part of the cylinders, which was the last part to be invaded by the fungus. It would therefore appear that the fungus consumed it as food. On this date, on each of the alkaline cylinders, the fungus had formed a stroma, which was gray-white, wrinkled, and mealy from the presence of great numbers of microconidia. In the tubes which received the most alkali the stroma covered about one-fourth of the surface and there were no perithecia. In those which received less alkali it covered about four-fifths of the cylinder in one tube, and only about one-twentieth in the other. In both of the latter perithecia were forming in small numbers, and were first visible on the fifteenth day. The tube containing the smaller stroma was now opened and the cylinder removed. It smelled like a soap barrel. The interior of the cylinder and of the stroma were alkaline, and also the beads of water excreted from it. There were very few perithecia. On the twenty-fifth day the remaining tube, which received the least alkali, was fruited rather sparingly, and the perithecia were red, but not ripe. In places the stroma was brown and destitute of perithecia; in other places it was grayish-white and destitute. In the tubes which received the most alkali the potato was now covered with a copious, puffed-up, tough, wrinkled, extensive stroma, which had become brown (color of of ascospores) since the last record. Some of this stroma was dug out and examined. The brown color was lodged in the walls of a closely felted mycelium. From this brown mycelium projected many short, colorless hyphæ, and these bore numerous non-septate, colorless, elliptical microconidia. Here and there were small masses of pseudo-parenchyma, probably the beginnings of perithecia, but nothing large enough to be distinguished by shape, even under the compound microscope, as perithecia by anyone not conversant with the method of origin of these bodies. In other words, the alkali did not hinder a copious development of stroma and an abundant formation of microconidia, but did hinder the growth of the perithecia. At the end of 47 days, during which there had been no development of perithecia in either of the tubes which received the larger quantity of alkali, one of them was opened and 4 c. c. of sterile 2 per cent malic acid water pipetted into it. At this time the potato was covered all over and hidden by a copious wrinkled stroma, which was brown on one side and about normal in color on the other. The next day (February 1) the acid was entirely neutralized, and the fluid around the cylinder still gave a distinctly alkaline reaction, although less than before. The mouth of the tube was now flamed and the fluid poured out. On February 5, as soon as it could be prepared, 5 c. c. of sterile 1 per cent malic acid water was added to this tube, i. e., enough to cover the cylinder. On testing with litmus after adding, the fluid gave a strong acid reaction. On February 6 there was no reaction to blue litmus, but a slight alkaline one with neutral litmus. An additional 5 c. c. of the 1 per cent malic acid water was therefore added, and in a few minutes bubbles of gas again began to arise. On February 7 the fluid was again tested, found plainly acid, and poured off. The cylinder was not, however, fully neutralized, for on February 10 the small amount of fluid in the bottom of the tube was again plainly alkaline. The cylinder was therefore covered once more with 6 c. c. of sterile 2 per cent malic acid water. This was poured off after 24 hours and the tube set away. Up to February 15 there were no perithecia visible in this tube, although it had been 62 days since the tube was inoculated and 5 days since the last of the alkali was removed. Ten days later, however, the stroma (both the brown and the yellow parts) was slightly frosted over with white hyphæ, recently formed, and bore several hundred pale coral-red perithecia, while many others were developing and were still white. The companion tube was free from perithecia and remained so until the end of the experiment (76 days). At this time the fungus in the tube which had received only one-half as much alkali had developed perithecia all over the potato, but they were still immature and either colorless, yellowish, or pale brown.

(3) In 40 days on rice with bicarbonate of soda there was no production of perithecia, although the tube had been copiously inoculated with ascospores. The tube contained 10 c. c. of rather dry boiled rice, to which was added 2 c. c. of a cold saturated solution of the sodium bicarbonate in water. In 5 days there was only a very feeble white growth. At the end of 40 days the fungus had grown sparingly over most of the rice grains in the upper half of the tube, and microconidia were present, but the rice had not changed color and no perithecia could be found.

OTHER BIOLOGICAL PECULIARITIES.

In addition to what has already been said, there are a few other peculiarities which may here be mentioned.

The fungus in all its varieties is a strict aerobe. Up to this time the writer has never been able to find any substance in the presence of which it will grow in the closed end of fermentation tubes. It seems entirely unable to obtain its respiratory oxygen from food substances.

The melon fungus is able to obtain its nitrogen from asparagin. In carbohydrate foods nearly destitute of nitrogen, and in which the fungus could make but a very feeble growth, it at once made an excellent growth on adding a small fragment of asparagin.

Old rice cultures of the melon fungus gave off a peculiar, rather pleasant, aromatic odor when boiled in water.

On boiled seeds of the cowpea the melon fungus grew vigorously and developed a peculiar musky, pungent odor, which was observed as soon as the tubes were unplugged and was quite unlike that just mentioned.

The melon fungus grew vigorously on potato and nutrient agar which was sterilized with a large quantity of sulphur dust.

The melon fungus grew on nutrient agar in the presence of large quantities of caustic lime and of carbonate of lime.

The melon fungus also grew (at first very slowly and in small patches) in a flask of boiled rice in the presence of very large quantities of sodium carbonate (one-half pound of rice cooked with about 300 c. c. of distilled water and 100 c. c. of a saturated solution of sodium carbonate, saturation temperature $25°$ C.). The quantity of alkali was sufficient to make the rice quite yellow.

In a test-tube culture, the cowpea fungus grew on a cylinder of potato which had been boiled with 2 c. c. distilled water and 1 c. c. saturated solution of sodium carbonate ($25°$ C.). After steaming, one-half of the fluid was poured off, so as to expose part of the potato to the air, and the surface was inoculated with ascospores.

Both the melon and the cowpea fungus grew in nutrient gelatin (+10, +20, 0, and −20 of Fuller's scale), but without liquefaction. The melon fungus grew rapidly on glycerin agar. In Dunham's solution with 2 per cent cane sugar and 0.2 per cent sodium carbonate the melon fungus made more growth than the cowpea fungus (ascospore strain). After 13 days the former filled the tubes (10 c. c. portions), while the latter had made only a trifling growth on the bottom of the tubes. In each the fungus was browned.

Both the melon and cowpea fungus made a very insignificant growth on dried figs steamed in distilled water, the very sweet substratum appearing to be unsuitable for growth.

In acid media, such as the juice of ripe Concord grapes, the growth of the melon fungus was much retarded, but it finally overcame the inhibiting substances.

The melon fungus grew well in sterilized horse dung and feebly on rotten wood (soft maple destroyed by wood fungi).

Bordeaux mixture sprayed upon young melon vines in no way checked the spread of the disease.

Carbonate of copper mixed with carbonate of lime and put into the hills at planting time, or soon after, did not protect either cotton or melon plants.

HOST PLANTS.

Neocosmospora occurs on cotton (*Gossypium herbaceum* and *G. Barbadense*), watermelon (*Citrullus vulgaris*), and on cowpea (*Vigna sinensis*). It probably occurs also on okra (*Hibiscus esculentus*), although the identification is not complete, depending solely on the character of the symptoms, on the presence of similar macro and microconidia, and on the occurrence of the disease in the same localities, no cultures or cross inoculations of the okra fungus having been made and no perithecial fruits having been discovered.

HABITAT, TIME OF OCCURRENCE.

This fungus lives from year to year in the soil. It is peculiarly a soil organism, always attacking the plant from the earth. The internal conidia (Pl. II, 8, 11; Pl. V, 4, 6) occur in the vessels of the living plant throughout the growing season, causing the disease known as "blight" or "wilt"; the external conidia (Pl. I, 9; Pl. II, 7; Pl. V, 3), whenever plants have been killed by the internal fungus; the perithecia, from August to November.

GEOGRAPHICAL DISTRIBUTION.

All stages of the fungus have been found by the writer at Monetta, S. C., on watermelons and cowpeas (1894 and 1895), at Charleston, S. C., on cotton and cowpeas (1895), and near Kingstree, S. C., on cotton (1895). The conidial stages have been found by the writer at Charleston, S. C., in okra and watermelon (1895), and at Chuckatuck, near Norfolk, Va., in watermelon (1898). Diseased watermelon plants containing the internal fungus were also received from Lykesland and Gurley, S. C. (1896), Pelham, Ga. (1895), and from Pepper Grove near Galveston, Tex. (1895). The cotton disease was also received from western Georgia in 1894. In addition, Professor Atkinson, in 1892, reported cotton disease associated with his *Fusarium vasinfectum* from seven localities in Alabama, and from Pine Bluff, Ark. The melon wilt was also reported to the writer in 1895 from Ocean Springs, Miss., by Prof. F. S. Earle, who stated that it had been in that vicinity for at least six years, and that in one instance it destroyed nearly an entire crop of melons. These melons were on land where melons formerly wilted to a slight extent and which had been in pasture for two years.

Finally, Mr. Wm. A. Orton found the cotton and cowpea disease at Dillon, S. C., in 1899. The cotton blight was pretty uniformly distributed over one 5-acre field. He also found it to a lesser degree in many other fields, and it was reported to be generally prevalent in that region.

This fungus, therefore, is widespread in the Southern States. It is to be looked for all the way from New Jersey to Texas, although it has not been definitely settled that it occurs north of southeastern Virginia or west of Arkansas. With exception of the one place in Arkansas it occurs, so far as known, only in the Atlantic Coast and Gulf States. It is specially to be searched for in New Jersey, Delaware, Maryland, and West Virginia, and in northeastern North Carolina, where a destructive watermelon disease of unknown origin was very prevalent about ten years ago. No account of this disease has come from any part of the Old World.

PARASITISM, INFECTION EXPERIMENTS, ETC.

This fungus is an active parasite and destroys a great many plants by first plugging the water ducts and afterwards invading the parenchymatic tissues. In case of the watermelon, the disease due to this fungus is so prevalent in places as to destroy large fields, and to threaten the extinction of melon growing for market purposes, notably at Monetta, S. C., Pelham, Ga., Chuckatuck, Va., and Galveston, Tex. The cotton fungus is also spreading from year to year, and has already spoiled many acres of valuable land on the fertile coast islands of South Carolina. Last year one large grower of sea-island cotton wrote that he had been compelled to stake out and abandon 15 per cent of his best cotton land, all of which is tile drained and under a high state of cultivation.

The fungus winters over in the soil and enters the plant through its underground parts. It first fills the vessels (Pl. I, *10*, Pl. IV, and Pl. V, *6*), causing, especially in melon vines, a sudden wilt of the foliage. Subsequently, as the plant dies, it invades all of the softer tissues and fruits on the surface (Pl. I, *11*, *12*, Pl. II, *6*). The internal fungus always precedes the external one, i. e., the external fungus has been found only on plants which have been killed by the internal one, and it almost always follows the latter, so regularly, indeed, as to be at once suspected of forming a part of the life cycle of the parasite. Several hundred plants each of melon, cotton, and cowpea in early stages of the disease, in different years and from different places, have been examined microscopically, and in every case the fungus was found plugging the vessels in quantity sufficient to account for the disease, and it had not yet invaded the parenchyma. In cowpeas, prior to their death, it is usual to find the fungus occupying the vessels of the xylem and browning their walls the whole length of the plant and not simply near the earth. In old, wilting melon vines (still alive and not showing any surface fungus), the internal fungus has been found in vessels of the stem 2 to 3 meters from the root, but, in the earliest stages of infection (the earliest yet observed), the fungus is found only in the vessels of the root and extreme base of the stem (hypocotyl). Many cotton, okra, melon, and cowpea plants further advanced in disease have also been examined microscopically. In the interior of these plants, both in the vessels and in parenchyma cells, there was an abundance of mycelium bearing great numbers of the small, colorless (white), elliptical microconidia, and at the same time, on the outside, on the salmon-colored conidia beds, enormous numbers of the big, lunulate, 3-to-5-septate macroconidia, the hyphæ of the fungus being readily traceable outward from the plugged or partially occupied vessels of the plant to the conidia beds on the surface. In early stages of the disease no external spores or hyphæ of this fungus are ever present, and the internal fungus is restricted to the vessels and connective tissue of the bundles, i. e., the non-lignified living parenchyma between the bundles is not yet invaded (Pl. I, 10 and Pl. IV). The arrangement of the external conidia beds in rows is due to the fact that the fungus comes to the surface of the dying plant along the lines of least resistance, that is, between the strands of thick-walled bast fibers. If the external fungus were an independent parasite or saprophyte due to aerial infections, the sporodochia would be scattered irregularly over the surface and not arranged in parallel rows, alternating with the strands of stereome, and always preceded by the internal fungus.

The same conclusion respecting the genetic relationship of the external and internal conidia has been reached as the result of plant-infection experiments, the watermelon being used for this purpose. More than one hundred and seventy-five typical cases of the melon wilt have been obtained in hothouses in Washington by simply burying a little of the melon fungus in the soil of the pots. On microscopic examination the fruiting fungus was found in the vessels of nearly every one of these plants (all examined) in such quantity as to readily explain the wilt. The soil and water were not sterilized, but under the circumstances

34

this was not necessary. The soil was from Washington where this disease does not occur, and melon plants grown in quantity in the same hothouses, in the same soils, and watered with the same water, subject, in a word, to the same conditions except as to inoculation, did not contract the disease. These experiments were in progress parts of two seasons. In many pots of the inoculated soil every plant contracted the disease within a few weeks (Pl. IX, 1). With two exceptions[1], not one case of the disease appeared in the uninoculated soils, although several hundred check plants were grown in the latter.

During the course of these inoculation experiments, which were performed in 1894 and 1895, it was found to be as easy to produce the melon wilt with mycelium derived from the external conidia (Pl. X) as with that derived from the internal conidia (Pl. VIII). The usual method of infection was by putting a little fragment of a pure culture of the fungus into the soil at some distance from the plant and at no great depth, so as to avoid breaking the roots. The wilt usually appeared suddenly in three to six weeks, whichever spore form was used for infection. The internal fungus in the vessels of the plant was the only one to be found at the time of the wilt, but later when the plant was dead the external conidia beds often appeared, and always if the air was not too dry. These infection experiments were repeated several times on a large scale with uniform results. Those with the two sorts of conidia were conducted in different greenhouses, with very great care, and, exclusive of the 4 cases already mentioned as occurring late in the course of two of the experiments, the numerous control plants in all cases remained healthy.

NUMBER OF SUCCESSFUL INFECTIONS.

Altogether more than 500 successful infections have been obtained with watermelon plants by simply inoculating the soil with the melon fungus, the above-mentioned 175 cases not including those obtained in 1896 and 1897 and described in the following section. All varieties of the watermelon appear to be susceptible. No attempts have been made to induce the disease by inserting the fungus directly into wounds, except a few futile ones made in the fields in South Carolina, when the disease was first discovered.

[1] (1) One case which appeared late in the course of the experiment (internal conidia, 1894) and the only plant out of many checks to contract the disease, although all were on the same bench and cockroaches ran about in the house.

(2) Three cases which appeared late in the course of the experiment (external conidia, 1895). These were in 2 pots only out of many checks which had been standing on the experimental bench for several months within 20 inches of numerous pots of inoculated soil, in each one of which several plants had wilted. Under the circumstances there can be no doubt that these were accidental infections derived from the inoculated soil or the wilted plants.

35

COTTON AND COWPEA INOCULATIONS, CROSS INOCULATIONS.

All of the cotton-plant inoculations have failed. These also were soil inoculations, and were performed on many small plants, using the cotton fungus, the cowpea fungus (cultures from ascospores), and the melon fungus. The experiments were repeated in different years and were continued in some cases for a long time with both sea island and upland cotton.

Melon-plant inoculations, on a large scale, also failed when the cotton fungus was used. These were soil inoculations. Pythium attacked and destroyed some of the plants and injured others, but no Fusarium disease appeared. The above statement should be qualified as follows: In the fall of 1894, 25 pots of melons were infected with the cotton fungus soon after the seeds were planted. For some weeks Pythium caused a good deal of trouble, and in one of the plants attacked by this fungus a small quantity of mycelium bearing the internal conidia of a Fusarium was found in the base of the stem. More cases, and typical ones, were anticipated, but they did not appear.

Cowpea inoculations also failed with both the melon and the cotton fungus. These were soil inoculations on plants in all stages of growth, from seedlings to plants two months old. Many plants were used, and the experiments were continued from four months to more than a year. The soil was copiously infected with the fungus, and repeatedly in some cases. Watermelons planted between the rows of cowpeas readily succumbed to the disease when the melon fungus was used. The above statement should be qualified as follows: On November 16, 1895, 182 cowpea seeds were planted in 26 pots of soil, in each one of which several melon plants had wilted the preceding June. The plants grew healthily all winter, but on February 14 the lower leaves of one plant began to wilt, and six days later the few remaining leaves fell off at a touch. The stem was green and appeared to be healthy except at the surface of the earth, where it was brown and partially rotted off, as when attacked by Thielaria basicola. In the upper part of the tap root, in the vicinity of this injury, the walls of some of the vessels were browned, and a few of these vessels were packed full of a fungus bearing the internal Fusarium conidia. More cases, therefore, and typical ones, were anticipated, but none appeared, although the plants were kept under observation for a long time.

Cowpea inoculations failed with the fungus derived from ascospores of the cowpea fungus. Nine large pots were used, but all of the 170 plants remained free from disease distinctly attributable to the Neocosmospora, although under observation for many months. These were soil inoculations, and an abundance of the fungus was used. A few melon plants grown in these same pots also remained free from disease.

Tomato plants in quantity were grown for a long time in soil full of the melon fungus without contracting any disease. This was done because a Fusarial disease of tomatoes occurs in Florida and also in

England.[1] Melon plants wilted readily in this soil. The tomatoes were developed in it from seedlings, and, to the number of more than 100, grew in the soil healthfully for several months (as long as the experiment continued) side by side with many watermelons, all of which contracted the disease and wilted.

Potatoes grown in soil full of the melon fungus also remained free from disease.

No inoculation experiments either on melons or cotton have been made with mycelium derived from the ascospores of the cotton or the watermelon fungus.

A detailed account of all of these experiments will not be given, but to show their extent and to re-enforce the above rather brief statement of the main results the following sample experiments are given:

(1) On September 5, 1894, 14 pots of good earth were planted with 56 cotton seeds and the soil of each one was infected with a pure culture of the melon fungus. During the fall and early winter the soil in these pots was reinfected three or four times with pure cultures of the melon fungus, but none of the cotton plants contracted the disease. On May 14. 1895, 46 cotton plants remained, and were healthy but for the attacks of red spiders. On this date watermelon seeds of three varieties were planted in each of the pots to see if the melon fungus was present in the soil. The seeds germinated well, and on May 25 each pot contained from two or three to six healthy melons. On June 7 the first case of melon wilt appeared—i. e., the cotyledons of a plant drooped in a suspicious way, and the conidia-bearing mycelium of the fungus was found in the vessels of the tap root. On June 8 the second case appeared. This plant was pulled and examined two days later, when the fungus was found plugging many of the vessels of the tap root. From this date cases became frequent. By July 9 the melon wilt had appeared in 10 of these pots. The cotton plants in these pots were still free from wilt and were making an increased growth, owing to the fact that the red spiders had been destroyed early in June by means of resin wash.

(2) On December 3, 1895, 28 pots of good earth were planted to sea-island cotton and an equal number to cowpeas. Each pot received 7 seeds. On December 5, before any of the plants were up, one-half the pots of each lot were copiously inoculated with pure cultures of the melon fungus, the whole of tubes 3, 4, 6, and 7. October 17 (rice cultures), being used for this purpose. The fungus was buried about 1 centimeter deep in the center of each pot. On December 11, as the plants were coming up, the other 14 pots of each series were also inoculated with the melon fungus. The pots of cowpeas received rice cultures 4, 5, and 8, October 8. The pots of cotton received rice culture 6, October 8, and 1, October 28. The fungus was buried in the soil the same way as before.

[1] Massee: The "sleepy disease" of tomatoes.—The Gardeners' Chronicle, Series III. Vol. XVII, 1895, p. 707.

On January 7, 1896, 15 pots of the cowpeas, which were now 6 to 8 inches high, were re-inoculated with the melon fungus. There were 89 plants in these 15 pots. The fungus was derived from 5 pure test tube cultures on bruised, steam-sterilized seeds of the cowpea, and was put into the top layers of the soil very copiously. The cultures were young (1 and 2, December 21; 10 and 11, December 16; and 7, December 31) and growing vigorously—i. e., each tube contained 10 to 15 cubic centimeters of the moist peas entirely overgrown and interwoven with the fungus. In these 56 pots at least 175 plants of cowpea and as many more of cotton were exposed to infection under conditions which appeared to be admirable.

This experiment was continued 14 months, during which time no positive results were obtained. None of the cotton plants contracted the disease and none of the cowpeas. In one only of the cowpea plants a very little of the mycelium of a Fusarium was found in a few of the vessels near the earth, but the symptoms were not typical for the cowpea disease, and the fungus may have been that of the non-parasitic Nectria mentioned below. On February 4, 1897, both varieties of cotton were still alive. A few of the cowpeas were still alive, but the majority ripened their seeds and died in the fall of 1896. At this time the stems of some of the latter bore sparingly and in a saprophytic manner (in no case high on the stems, but always near the moist earth), the pinkish conidia beds of a Fusarium. These first began to appear in November—i. e., 11 months after the soil was inoculated. Some of these compact sporodochia may have been the conidia beds of the melon fungus, but they did not appear to be parasitic—i. e., they were not preceded by an extensive occupation of the vessels of the plant and were not arranged in rows up and down the stem,[1] but rather were clustered on the moist bases of some of the decaying stems. Moreover, on one stem they appeared in connection with the red perithecia of a Nectria. These perithecia in shape and color much resembled those described in this bulletin, but the ascospores were colorless, thin-walled, elliptical-pointed, smooth, 1-septate, and 8 to the ascus—i. e., typical Nectria sporidia.

To determine whether the melon fungus was alive in the soil, the earth was knocked out of the pots and the cotton and cowpea stems buried in it to form the substratum of a bed on one of the hot-house benches. This earth was then covered with three pailfuls of clean sand, on top of which an inch of good potting soil was spread. On February 10 this bed was planted with the seeds of two varieties of watermelons. Seeds enough for 500 plants were put into the earth, but

---

[1] This fungus did not in any case fill the vessels and brown their walls the whole length of the plant, as the cowpea fungus does, nor even for a few inches. Neither did the external conidia beds appear in the proper way. In the cowpea disease these external conidia beds occur by thousands on the dry stems at all heights from the surface of the earth to the top of the plant (3 feet or more), and are arranged very regularly in parallel rows lengthwise of the stem. (Pl. I, *II*, *I2*.)

they came up badly. On March 1, 75 seedlings of the Rattlesnake melon were up and 42 seedlings of the Seminole. On March 5 two plants showed drooping cotyledons, and on making sections of the tap root the melon fungus was found fruiting in the vessels of each. During the next two weeks 24 melon plants wilted, and the characteristic fruiting fungus was found plugging the vessels of the taproot or hypocotyl of each one. Subsequently many additional plants were attacked and destroyed, showing clearly that the melon fungus must have been alive in the soil during the whole time that the cotton and cowpeas grew in it unmolested. The presence of the typical conidia-bearing fungus in the vessels of the wilting plants was determined, in nearly all cases, by a microscopic examination, although this became rather monotonous toward the close of the experiment. It was on this bed that the tomatoes grew unmolested from February to midsummer, and this, too, although many of the wilted melon plants were buried from time to time close to the roots of these tomatoes. Twenty cowpea plants which came up by accident (from seeds in the buried refuse) also remained healthy until they were removed in July.

(3) On April 12, 1897, another attempt (the fifth or sixth) was made to infect cotton and cowpeas with the melon fungus. For this purpose in one of the hothouses there was prepared a bed of good well-rotted potting earth. This bed was 6 feet long, 3 feet wide, and about 8 inches deep. As soon as this bed was ready, cultures of the melon fungus were buried in it, in rows, at right angles to the long axis of the bed, and 4 inches apart. Twenty-six test-tube cultures of the fungus were used for this purpose. These were vigorous growths on slant agar, potato, etc., which had been started for this purpose April 6. Grooves about an inch deep were made in the soil. The fungus was then uniformly distributed in these furrows and covered with the loose earth.

On April 13, 400 seeds of the Georgia Rattlesnake melon were planted in this bed, in 15 rows, alternating with the rows of the buried fungus, and consequently only 2 inches either way from the fungous masses. The germinating capacity of these seeds had already been tested and found to be high, but in this bed only about half of them came up.

On April 29 about 200 melon plants had come up, and none of them were unhealthy. The first case of melon wilt appeared April 30 and the second May 3.

On May 3—i. e., as soon as it had become apparent that the melon fungus was active in the soil—200 seeds of cowpea and 175 seeds of sea-island cotton were planted between the rows of melons—i. e., on top of the rows of buried fungus. To help on the disease and add interest to the experiment, the spores of *Thielavia basicola* were also planted in the bed in one or two places. This fungus makes wounds in cotton and cowpea plants at or beneath the surface of the earth from the surface inward, and it was thought that possibly such injuries might favor the entrance of the Fusarium.

On May 8 there were 6 additional cases of melon wilt with fruiting Fusarium in the vessels of each one, and one melon plant had succumbed to a combination of Thielavia and nematodes.

On May 10 the cotton and cowpeas were coming up nicely, and more of the melons had wilted. On May 21 there were 65 wilting melon vines pretty uniformly distributed over the bed. The cotton and cowpeas growing between these rows of melons showed no signs of the wilt disease. A few of them had rotted off at the base with Thielavia, but no Fusarium was associated with it. At this time the cowpeas were 6 inches high with the first trifoliate leaf coming, and the cotton plants were about 4 inches high with the first true leaf coming. By May 25 melons enough had wilted to make a total of about 100 cases. The cotton and cowpeas were free from wilt and growing very satisfactorily, considering how closely they were planted. By June 1 there had been 166 cases of melon wilt on this bed. All of these plants were examined microscopically, and in every one there was an abundance of the Fusarium in the vessels of the taproot or stem, or both. In the recently wilted plants the fungus was restricted to the vessels; in those which had been wilted some days, but were not yet dry-shriveled, it was also in the parenchyma, but had not reached the surface. Only 16 healthy melons remained, and these subsequently contracted the disease and died.

In digging and pulling out the diseased melons for examination the roots of the cotton and cowpea plants must have been considerably broken and disturbed, but neither in this way nor by the aid of the Thielavia or of another fungus destitute of fruit, but resembling a Pythium, was the melon fungus able to find its way into the cotton or cowpea stems.

Up to August 8, when the experiment was discontinued, all of the cotton and cowpea plants remained free from this disease. At this time they were large plants, and the cotton was suffering considerably from crowding.

ONE FUNGUS, OR THREE?

It is my intention to repeat and extend the ascospore inoculation experiments as soon as time permits, and also to settle more definitely by means of additional cross inoculations whether under any circumstances the fungus from any one of these plants will ever transmit disease to the others. Much time has already been devoted to this problem, which is one of great practical importance. From certain cultural peculiarities of the fungi, from the uniformly negative results of the hothouse experiments and from certain field observations, it now looks as though these were separate diseases due to closely related but not identical organisms. This is also the opinion of some very well-informed growers.

One of these field observations may here be given. In July, 1894, the writer examined a large field of upland cotton belonging to Mr. T. S.

Williams, of Monetta, S. C., without finding any cases of the cotton wilt, and Mr. Williams stated that there had never been any in it. This cotton was planted on land where watermelons had wilted badly in 1893. It was therefore on a large scale an experiment similar to those described above. In 1895 the field was again planted to cotton, alternating with cowpeas—i. e., 3 rows of cotton and 3 rows of cowpeas. In September of that year the writer again examined the field quite carefully, and with the same negative result. No cases of cotton wilt were to be found, but there were hundreds of wilting and dead cowpea vines, and the latter bore the external conidia beds of the lunulate-spored Fusarium quite regularly, and these were arranged in the manner shown in Pl. 1, *11, 12*. The melon fungus was undoubtedly present in the soil, but there were no melon vines on the field by means of which to establish this fact conclusively. At my request Mr. Orton reexamined this district in the summer of 1899. He found an abundance of melon and cowpea wilt, but none of the cotton disease.

Morphologically, on the contrary, so far as I have been able to determine from the careful examination of a great many specimens, this fungus is specifically identical on all of these plants, the only doubt being whether it may not have varied enough physiologically not to be transmissible from one host to the other. There are slight variations in the length of the beak and in the size and color of the perithecia on the different host plants, but these are not constant. The ascospore variations are also not constant. As already stated, the ascospores of the form on watermelon are usually smaller and more decidedly elliptical, and a larger proportion are smooth (perhaps because small). When this was discovered an earnest attempt was made to distinguish two or more species, but further studies developed the existence of all sorts of intermediate forms and sizes and left no morphological standing ground for any such conception. For example, perithecia were found on the watermelon at Monetta, S. C., in September, 1895, which contained globose wrinkled ascospores $12\,\mu$ in diameter; ascospores 12 by $13\,\mu$ and 12 by $14\,\mu$ were also observed. Other perithecia on the same root bore elliptical smooth spores. Indeed, ascospores from the same perithecium may be globose or elliptical, wrinkled or smooth. Red perithecia taken from the roots of wilted, dead cowpeas, at Monetta, September 10, 1895, were indistinguishable in color, size, and shape from those found on watermelon roots, and bore ascospores of the following sizes: 9 by 10, 9 by 14, 10 by 10, 10 by 12 (wrinkled), 10 by 13, 10 by 14 (smooth), 12 by 12 (wrinkled). Ascospores from perithecia which grew on the roots of sea-island cotton on James Island were 10 by 10 (smooth and not many so small), 10 by 12 (smooth and wrinkled), 12 by 12 (wrinkled), 13 by 14 (wrinkled). Owing to variability, it is likewise impossible to distinguish the fungus from the various hosts either by means of the mycelium or by the external or internal conidia. At most the differences can be scarcely more than varietal—i. e., such as might be induced by the long-continued use of different substrata. This is true even if the fungus can not be transmitted from one host to another, and a sufficient number of experiments have not yet been made to enable one to declare without reserve that this never takes place under any circumstances—e. g., with help of some other fungus, or under peculiar conditions of environment not yet discovered. At present, therefore, nothing remains but to consider the fungus as one species and to record the forms on the other host plants as varieties. Possibly it may finally turn out that they do not deserve even this rank, but the writer does not now feel justified in giving them any less.

VITALITY.

From its vitality under adverse conditions, its ability to live in the dung heap and in the soil, and the ease with which it may be cultivated on all sorts of artificial media in the laboratory, this fungus must be regarded as a serious enemy to agriculture. While we know it only on the plants mentioned, it is probably capable of attacking other species, and ought certainly to be expected and looked for on other hosts.[1]

The length of time the fungus will remain alive in the earth is remarkable, and adds greatly to the difficulty of combating it. Why it should ever disappear, any more than a bad weed, when once established in a cultivated soil, is not clear. It should certainly be regarded as a weed, and one the eradication of which presents unusual difficulties. In extensive field culture it has been found unsafe to plant lands which have once suffered from it until after a lapse of several years—five to seven, according to south Georgia melon growers, and certainly more than three, as shown by a 7-acre field test in South Carolina, which came under my own observation in 1895. Only 3 wagonloads of melons were obtained from the whole field.

The melon fungus has lived a year in the dried-out soil of pots used in my greenhouse experiments, and a very similar fungus parasitic on cabbage remained alive in dry soil three and one-half years.[2] The

---

[1] Since this was written Mr. Orton has found the fungus on James Island, South Carolina, in a weed, the Cassia obtusifolia L. Not many plants were attacked, but the external symptoms were identical with those on the cotton and cowpea, while the walls of the vessels of stem and root were stained brown and the lumen was filled frequently with mycelium, abstricting the typical microconidia, and sometimes browned, as in cotton.

[2] The writer has just concluded an experiment with a similar looking and acting cabbage Fusarium. In parts of New York, Virginia, and Maryland this fungus has troubled the market gardeners, in some cases rendering impossible the profitable culture of cabbages on large and fertile fields—e. g., a field in New York which formerly yielded from 90,000 to 95,000 heads of marketable cabbage for each 100,000 plants set out, can now be depended upon for only about 30,000 heads; on a field in Maryland which formerly yielded good crops, the cabbages were so badly affected this year that the ground was replowed and planted to other crops in the middle of the season. The symptoms in the cabbage are slow growth, refusal of the heads to form, a sickly color, and the premature shedding of the lower leaves, from the axils

melon fungus lived in a dried-out condition in one of my agar tube cultures three and one-half years; in another it was alive at the end of ten months and twenty-three days; in a third trial it was found alive in 7 out of 8 test-tube cultures which had been in a warm, dry place in a dried-out condition for nearly two years. These last tubes were inoculated at various dates between July 26 and October 8, 1894, and the test was made January 19, 1897. The ascospores of the cowpea fungus remained alive in a dry condition 16 months.[1]

SYMPTOMS PRODUCED. ETC.

The gross symptoms in the watermelon are sufficiently shown on the accompanying plates (Pls. VII, VIII, X). They are those of a plant

of which short sprouts frequently push out. This disease was first studied by the writer in 1895. It attacks the plant in the same way as the melon fungus—i. e.. from the soil—and destroys it by plugging the water ducts. It first produces in the vessels of the living plant great numbers of microconidia (8 to 13 by 2 to 4 μ), and then macroconidia on the surface, in the same way as the melon fungus. In July, 1895, a quantity of soil was obtained from one of these badly infested fields (near Albany, N. Y.) and was stored away in a dry basement for three and one-half years. It was then removed from its original packings, put into a clean pine box made from freshly planed lumber. planted with three varieties of cabbage, placed in an upper window of an unoccupied laboratory room, and watered with distilled water. Some of the plants were attacked by Pythium and others by this Fusarium, which was found fruiting in the vessels. The checks did not contract the disease. The conditions under which the experiment was made point unmistakably to the soil as the source of the infection in case of these plants. This experiment, in connection with those which have been made on the watermelon, renders it probable that all parasitic soil Fusaria are alike in being very resistant to drought and other conditions unfavorable to vegetative life.

Embolisms of the vascular system due to fungi of the form-genus Fusarium are now known to the writer in plants of many different families. (See partial list on p. 43.) Most if not all of these fungi enter the plant from the earth. The fungi occurs so regularly in connection with these diseases that we are warranted in assuming them to be parasitic, although as yet in most cases no infection experiments have been instituted or brought to a successful conclusion. Judging from the results obtained with the watermelon and cabbage, it appears extremely probable that in the Cephalosporium and Fusarium stages of a variety of Nectriaceous fungi we have to deal with a large group of destructive soil parasites the very existence of which, in the earth, as active parasites, was not suspected until very recently—i. e., until the announcement of my results in 1894. (Am. Asso. Adv. Sci.)

[1] Some additional tests were begun September 28, 1899, as this bulletin was passing through the press. The watermelon fungus was found to be dead in each one of the 10 bread cultures already mentioned (p. 16) and in a few other cultures of the same age (hominy, cowpeas), but was still alive in numerous test-tube cultures of horse dung. These were made October 27. 1894 (from internal and external conidia), and have been dried out for at least four and one-half years. Twenty-two tubes were tested, and the fungus was found alive in each one.

Four test-tube cultures of the cotton fungus (from internal conidia) were also tried at the same time, and the fungus was found to be alive in each one. These 4 cultures were 5 years old. They were made on sterilized stems of the watermelon, October 3, 1894, and have been dried out and in a dry laboratory for at least four and one-half years.

The cowpea fungus (ascospore strain) was dead in each one of the 12 banana cultures made December 6, 1895.

transpiring freely and insufficiently supplied with water, although at the same time there is an abundance of moisture in the soil. The uniformity with which this fungus first seeks out the vessels of the plant is very striking (Pl. I, 10, Pl. IV, and Pl. V, 6). This explains the sudden wilt of the foliage. The water ducts are clogged to such an extent that they can not function. That this wilt is attributable to lack of transpiration water, brought about by partial or complete clogging of the vessels, is shown by the fact that large plants which have begun to wilt frequently recover for a day or two if a rain sets in and the air continues moist. During such weather the progress of the disease in old vines is almost at a standstill, but it recommences when the sun shines out and the moisture of the air is dissipated. The mechanical nature of this obstruction is also shown by the fact that collapsed leaves frequently recover their turgor on cutting stems above the fungous plugs and plunging them into water, e. g., on cutting across the second internode when the plugs are confined to the hypocotyl and taproot. It is also shown by the fact that the terminal portion of shoots which have collapsed and the basal vessels of which are plugged by the fungus (as shown by the death of all the lower leaves and of all the other branches) may survive for several days, if the weather is not too hot and dry, provided they are attached to a large melon from which they can draw a certain amount of water (Pl. VII, 2, right-hand branch). Finally, it is much easier for the fungus to plug all of the vessels of a small plant than of a large one, since in the taproot it has to make but a little growth through tender tissues to accomplish this. This explains why *young vines* frequently wilt even during rains or when the soil and air are very moist. The leaves of the cowpea usually unjoint and fall off, leaving the green stems bare. Often some of them become yellow and fall without first showing signs of wilt, as Professor Atkinson has recorded in case of the cotton disease. The watermelon leaves do not become yellow or detach, but wilt suddenly in large numbers and shrivel, so that a large, healthy-looking vine may lose all of its foliage in twenty-four to forty-eight hours (contrast Pl. VI and Pl. VII). The cotton plant is less susceptible, contains, as a rule, less of the fungus, and often recovers partially, so as to produce some fruit. In such instances the fruit-bearing stalks push out of the base of the stem and finally hide more or less completely the main shoot which has been killed by the wilt.

The xylem of the diseased plants always becomes brown (Pl. I, 10 a, b), and in case of the translucent stem of the cowpea this stain shows through the green bark, giving an unusually dark appearance to the still living stem. This browning of the xylem appears to be common to all plants attacked by parasitic soil Fusaria, e. g., cotton, okra, cowpea, beans, watermelon, squash, potato, tomato, eggplant, red pepper, sweet potato, cabbage, carnation, asparagus, pineapple, and others. This browning begins in the walls of the larger vessels, and often it is confined quite exclusively to the xylem for a long time, the pith, bark, and phloëm remaining free from stain. In the cotton fungus the older mycelium itself, inside the vessels, is frequently stained yellow or brown. Browning of the mycelium of the melon fungus inside the vessels of the plant has not been observed, but it occurs in pure cultures on boiled melon stems, etc. Once started, in cotton at least, this browning may extend long distances through the woody parts of the stem with very little fungus to help it on.

OTHER WILT DISEASES.

This disease should not be confused with the cotton-root rot of Texas, from which it appears to be distinct,[1] nor with another wilt of cotton, cowpeas, etc., common in parts of Florida and first described by Prof. P. H. Rolfs. This is associated with a fungus which attacks many kinds of plants, wild and cultivated, enveloping the base of the stem externally with a copious, white, rather straight mycelium, bearing on its surface large numbers of small sclerotia which are first white, then fulvous, and finally dark brown and smooth. When grown on nutrient media (agar), the fungus reproduces itself by another crop of sclerotia, and so on indefinitely (a year in my laboratory). These sclerotia are mostly 0.8 to 0.9 by 1.2 to 1.3 mm.

RELATIONSHIPS.

The perithecium of Neocosmospora much resembles a medium-sized bright red Nectria, and if gathered in an immature condition would naturally be placed under Nectriella, as the spores are smooth and colorless and there is no indication of any septum, even in the most elongated or in very old spores. When ripe, however, the ascospores are distinctly brown, and the fungus clearly belongs to a new genus. It possesses some of the characters of both Nectriella and Melanospora, but is distinct from either. On the whole it seems to be nearer to Nectriella than to Melanospora, although the spores are brown. It is not very closely related to Melanospora as originally established by Corda and defined in Saccardo's Sylloge Fungorum. It is widely different from *Melanospora chionea, M. leucotricha, M. zamiæ, M. vitria, M. zobelii, M. coprina, M. lagenaria, M. parasitica,* and all other Melanosporas and Sphærodermas which the writer has been able to examine or to find figured. The ascospores of this fungus are particularly unlike those of Melanospora. In the latter they are smooth, often lemon-shaped, or even apiculate, and frequently oblique or flat on one or both sides. The general appearance of Melanospora indicates that it might perhaps be properly excluded altogether from the Nectriaceous fungi, its affinities

[1] Four hundred cotton plants killed by this disease were received from two different localities in Texas in 1895. Pammel's Ozonium was present, but not a trace of any stage of this fungus could be found. These plants came from regions much subject to root rot.

being with the Sordariaceæ, as Schröter pointed out in his Kryptoga-
men-Flora von Schlesiens. The development in the cowpea fungus of
perithecia with and without beaks on mycelium derived from a single
ascospore makes it probable that the same thing occurs in other genera
and tends to confirm Winter's view of the untenability of the genus
Sphæroderma, which was erected to include those forms of Melanospora
destitute of a beak. The development of the perithecia of this fungus,
either in the soil near the host or on the surface of the latter (the more
common way), or under the substratum in rifts or cavities of the host,
shows that this character has no particular value and leads one to suspect that the genus Hyponectria may also have no sound physiological
or morphological basis.

This fungus is most nearly related to the Cosmospora of Raben-
horst, which is a good genus and should be reestablished. It differs
from Cosmospora chiefly in having non-septate ascospores and a
wrinkled exospore, the former having 1-septate ascospores with a papillate or verrucose exospore. My cultures have shown beyond doubt
that the ascospore is less readily modified by the substratum than any
other part of the fungus, and therefore the most important for purposes
of classification.[2]

---

[2] These two genera may be defined briefly as follows:

Cosmospora Rabh. (emend.).
 Perithecia as in Nectria (red in the known species). Asci numerous. Ascospores
 8 in one row, brown, oblong elliptical, 1-septate, usually more or less constricted
 at the septum, with a distinct papillate or verrucose epispore. Paraphyses
 present, inconspicuous, broad, loosely jointed, unbranched. Conidial stages
 unknown.
  1. C. coccinea Rabh.
  2. C. Cameroensis (Rehm).
   Syn. Sphæroderma Cameroensis Rehm.
Neocosmospora.
 Perithecia as in Nectria (bright red in the known species). Asci numerous. Ascospores 8 in one row, brown, globose or short elliptical, continuous, with a distinct wrinkled epispore (the latter sometimes wanting in the smaller spores).
 Paraphyses present, inconspicuous, broad, loosely jointed, unbranched, consisting of about 5 cells. Three conidial stages, viz, Cephalosporium, Fusarium, and Oidium.
  1. N. vasinfecta (Atk.). On cotton (parasitic).
    Perithecia as below. Spores usually globose, wrinkled, generally about
    10 by 10 μ.
  α. Var. tracheiphila (Smith). On cowpea (parasitic).
    Perithecia quite variable, but mostly 250 to 350 μ tall by 200 to 300 μ
    broad, with or without a short neck; on the dead roots or in the soil
    over them. Spores usually globose, wrinkled, mostly 12 by 12 μ.
  β. Var. nivea (Smith). On watermelon. Very actively parasitic. Enters
    the plant from the earth and plugs the vascular system, causing a sud-
    den extensive wilt of the foliage.
    Perithecia as above. Spores globose or elliptical, wrinkled or smooth,
    generally smaller than in the preceding and more often elliptical, but
    variable.

SCIENTIFIC NAME.

For the present, at least, this fungus may be designated as follows:

NEOCOSMOSPORA VASINFECTA (Atk.)

Syn. *Fusarium vasinfectum* Atk.

On cotton. Probably also on okra. Parasitism not proved. Genetic connection of various spore forms not proved. Chlamydospores not observed.

α. Var. tracheiphila.

Syn. *Nectriella tracheiphila*, Erw. Sm.

On cowpea. Parasitism not proved. Genetic relationship of ascospores, macroconidia and microconidia established. Chlamydospores not observed.

β. Var. nivea.

Syn. *Fusarium niveum*, Erw. Sm.

On watermelon. Not known to occur on other Cucurbitaceæ.[1] Parasitism fully established. One of the most destructive soil parasites known. Genetic connection of conidial stages fully established. Perithecia identical with the preceding have been found on stems killed by the internal fungus, but the genetic relationship has not been proved by exact culture experiments.

By "not proved" the writer does not mean that he has himself any doubt whatever as to the parasitic nature of the cotton or cowpea fungus or as to the genetic relationship of the various spore forms on cotton and of the perithecia to the conidial stages on the watermelon, but that these points have not been *definitely settled* by satisfactory infection experiments and by deriving one spore form from the other in pure cultures. Without such experiments the proof remains incomplete. The field evidence, however, of the parasitic nature of the cotton and cowpea fungus is of the most convincing sort, i. e., the fungus is always present in the vessels of the diseased plants, nothing else is always present, and the disease occurs year after year on the same soils. This constant association of the fungus with the cotton and cowpea disease makes it reasonably certain (in the light of what has been accomplished with the watermelon fungus) that abundant infections will be secured at no very distant day. The failures thus far have probably resulted from a greater resistance on the part of cotton and cowpea, or from the fact that the natural method of infection has not been discovered. Possibly the cotton and cowpea may be subject to infection only during germination or early stages of growth, or only

---

[1] A destructive disease of cucumbers and muskmelons associated with a Fusarium in the stem bundles has been reported from Ohio by Prof. A. D. Selby. The occurrence of a similar disease in muskmelons has just been reported from Connecticut by Prof. William C. Sturgis. A Fusarial disease of squashes is known to the writer. Possibly these diseases are due to the watermelon fungus, but the writer has not observed diseases of other Cucurbitaceous plants in regions subject to the watermelon wilt, and until cross-inoculation experiments have shown their identity it is proper to consider these diseases as distinct.

when injured in some particular way. If these plants are infected chiefly during the seedling stage, then the growth of young plants in uninfected seed beds and their transplantation to infected soils only when they have passed out of the receptive stage would afford relief. This is offered, however, only as a working hypothesis. All we yet know is that frequently many plants of cotton and cowpea on infected land come to a healthy maturity, i. e., there is not such a sweeping general infection as in case of the watermelon. Possibly, however, this is a statement depending on an insufficient number of observations. It is true so far as my observations have extended.[1]

Melon plants grown for several weeks in good earth, i. e., plants with one true leaf and the second one beginning to develop, are still freely subject to soil infection. The shortest period of incubation observed by the writer has been about 6 days, i. e., the cotyledons wilted, as shown in Pl. VIII, the first or second day after the plants emerged from the ground. The longest period of incubation, or rather lapse of time between infection of the soil and appearance of the disease, was 81 days. In this case the seed was planted April 25 and the soil of the pot infected April 25 and the vine showed no symptoms of disease until July 15, when it suddenly wilted. Thirty-one other vines of the same series contracted the disease between June 1 and July 11. The soil of these pots was reinfected May 20, no cases having appeared. In both instances the fungus used was derived from a big, arcuate, several-septate, external conidium. Vine 32, which wilted at the end of 81 days, was 4 feet long. There was an abundance of fungus in the vessels and it bore only the small, colorless, elliptical microconidia; these were non-septate, straight or slightly curved, and measured 6.5 to 24 by 2 to 4 μ. In field culture a period of 80 days frequently intervenes between planting and the first appearance of disease in a given vine.

The question of parasitism of the cotton fungus was left an open one by Professor Atkinson, as shown by the following citations:

"The Fusarium was considered not to be a sufficiently aggressive parasite to be able to make its way into the ducts of the circulatory system unaided." It is suggested that a damping-off fungus may first open the way. The only two infections obtained by him with the cotton Fusarium were on plants with open wounds made by the "sore-shin" fungus. "The discoloration and disease of the ducts is started

---

[1] Mr. Orton observed in 1899 in various parts of South Carolina what had previously escaped my attention, viz, that every cotton plant in the vicinity of dead and dying plants is dwarfed and does not bear as much fruit as it should, even when it shows no symptoms of the disease. This dwarfing is most conspicuous toward the end of the growing season and is associated with the presence of the fungus in the vascular system of the side roots of the plant, the vessels of the tap-root and of the stem being free or very nearly free from the fungus. This may help to explain the statement often made by cotton growers that one stalk in a hill will blight and another will not. In reality, it may be only a question of degree, all being more or less affected.

by the injury from the 'damping-off' fungus." "While I do not wish to be understood as making any positive assertion in favor of the Fusarium being the cause instead of bacteria [which were sometimes found associated with it]. I do think the evidence thus far in hand gives greater support to the former view. The Fusarium is invariably found both in cotton and in okra afflicted with the disease. Bacteria are not always found in the diseased tissues, etc." Some experiments by Professor Atkinson also led him to believe the external Fusarium on cotton distinct from the internal one.

RESTRICTION OF THE MELON WILT.

While no cure is known for this disease, our knowledge of its cause and manner of spread is now sufficiently exact and complete so that certain rules of practice may be given. By carefully following these the farmer will frequently avoid very serious losses.

(1) Fields already infested with this fungus must not be planted to melons for a long series of years.

So far as yet known, canteloupes, cotton, peanuts, cowpeas, soy beans, or velvet beans (Mucuna sp.) may be planted on such fields without danger.

(2) Fields free from this disease may become infected by the wash from lands already infested, and probably, also, by means of the dirt adhering to agricultural implements and to the feet of horses and cattle. For this reason, if cattle are pastured on such fields, they should not be allowed to roam freely over uninfected parts of the farm, and tools used on such lands should at least be scoured bright before using on other fields. Where uplands are infected the wash from sudden heavy rain storms should be turned aside, as far as possible, from uninfected lowlands.

(3) Inasmuch as the vitality of the fungus is great and the wilting melon vines are full of it, the danger for a half year or more from such vines is very great. All such plants are magazines of infection. They should be pulled while green, stacked with brush, and burned. Large growers of melons could well afford to keep one man in the field all the time for this purpose. The plants should be removed as soon as they show distinct symptoms of the wilt, because at this time the fungus is still confined to the interior of the stems and not likely to be scattered about by the removal, as would be the case a few weeks later when the vines are dry and the fungus has fruited abundantly on their surface. This precaution should not be neglected simply because fields show only here and there a wilted vine, since in course of a few years such fields have been known to become so thoroughly occupied by the fungus as to altogether prevent melon growing.

(4) Occasionally the fungus is introduced into the barnyard, so that the dung pile becomes a source of general infection to fields previously free from the disease. This is apt to be the case where "melon hay" is

fed or used for bedding.[1] The writer discovered one striking outbreak of this disease in South Carolina in 1894 which could be accounted for in no other way, the disease making almost a clean sweep on the five acres which received most of the manure. If there is the least reason to suspect the manure pile, commercial fertilizers should be used instead.

(5) Farmers whose lands have become generally infected are advised to grow other crops on their own fields, and to rent uninfected land from their neighbors for the purpose of melon growing.[2]

It is a source of regret that this bulletin could not be made more complete, but during the period covered by this investigation many other important lines of work have also demanded attention. In the scant time at the writer's disposal for this special work he has therefore done as much and as well as he could, and must so leave it.

So far as relates to the life history of the fungus, the most important things remaining to be done are as follows:

(1) Determine time and manner of infection of cotton and cowpea plants.

(2) On cotton establish the genetic relationship of the various spore forms, perithecia and external and internal conidia.

(3) On watermelon obtain experimental proof that the perithecia on the dead stems bear the same relation to the internal and external conidia as they do in case of the cowpea fungus.

(4) Obtain infections with ascospores on cotton, melon, and cowpea.

(5) Connect the okra fungus with the cotton fungus by experimental studies.

(6) Determine why it is impossible to grow perithecia from some conidia and easy to grow them from others. Does remoteness of origin from the ascospore interfere?

(7) Determine by additional cross-inoculations and field observations whether under any circumstances the fungus from one host plant can transmit the disease to another host plant.

(8) Determine how and where the melon fungus gains an entrance into the plant.

(9) The ease with which the perithecia may be grown would seem to offer a good opportunity for studying the question of the sexual or non-sexual origin of the ascospore fructification in this group.

---

[1] Melon hay consists of hay made on melon fields late in the season, after the melon crop has been harvested. It is composed of wild grasses interspersed with dead melon stems. If the latter are infested with this fungus, then the dung heap becomes inoculated, and subsequently any land on which this manure is used.

[2] This advice was given by the writer to Mr. T. S. Williams, of Monetta, S. C., in 1894, with the happiest results, as indicated by the following excerpt from an unsolicited letter written by him in 1897: "Financially your researches here have been worth thousands of dollars to myself and others. By the information you gave us we were able in a great measure to avoid land likely to die."

Many practical questions are left for further investigation, especially all that relates to cotton.

EXSICCATI.

Specimens of this fungus will be distributed in Ellis and Everhart's Fungi Columbiana (Century 15). These sets include only the fungus as it occurs on cowpea and on culture media derived from this source. These specimens will consist of the following material: (*a*) Perithecia on stems and roots of the cowpea (*Vigna sinensis*); (*b*) pure cultures of immature perithecia grown in the laboratory from ascospores sown on steamed sterile potato; (*c*) ripe perithecia from pure cultures on steamed potato; (*d*) stems of cowpea with the vascular system plugged by the white fungus, which bears microconidia (these stems were green when gathered and bore no external conidia beds, but in many instances, to my chagrin, the fungus pushed through and fruited on the surface while the stems were drying); (*e*) external salmon-colored conidia beds (macroconidia) on stems previously killed by the internal fungus and which were dry when gathered. Type specimens have been deposited in the cryptogamic herbarium of the Division of Vegetable Physiology and Pathology, United States Department of Agriculture, and these specimens, which have been selected with equal care and are of equal value, may be regarded as co-types.

PREVIOUS LITERATURE.

(1) Atkinson. *Fusarium vasinfectum*. Some Diseases of Cotton. Bull. No. 41, Dec., 1892, Agricultural Experiment Station, Auburn, Alabama, pp. 19 to 29, with 3 figures: a diseased leaf, internal mycelium in vessels of cotton plant, mycelium and conidia from cultures.

(2) Atkinson. Diseases of Cotton in "The Cotton Plant," a Bulletin (No. 33) issued by the Office of Experiment Stations, Dep. of Agric., Washington, D. C., 1896, pp. 287-292. Nothing new added.

(3) Smith. *Fusarium niveum*. The Watermelon Disease of the South. Proc. Am. Asso. Adv. Science, 1894, p. 289.

(4) Smith. *Nectriella tracheiphila*. The Watermelon Wilt and other Wilt Diseases due to Fusarium. Proc. Am. Asso. Adv. Science, 1895, p. 190.

(5) Smith. The path of the water current in cucumber plants: (3) The result of parasitic plugging of the vessels. American Naturalist, 1896, p. 561.

(6) Smith. The Spread of Plant Diseases: A consideration of some of the ways in which parasitic organisms are disseminated. Tr. Mass. Hort. Soc., 1897.

A fungus or supposed fungus, described very imperfectly by L. v. Schweinitz in his Synopsis fungorum Carolinae superioris, and by Fries in his Systema Mycologicum, as *Sphæria gossypii*, has caused the writer some trouble and should be mentioned here. It was found on unripe cotton bolls and Fries stated that he had seen dried specimens. From the description in his Systema, making allowance for inexact observations and unwarranted inferences, it seemed at first not unlikely that the perithecia found by me on the cotton plant might be the old *Sphæria gossypii* of de Schweinitz. This view I have now abandoned as untenable.

The origin and history of this name, which must be regarded as a *nomen excludendum*, are as follows:

(1) De SCHWEINITZ. Synopsis Fungorum Carolinae superioris, p. 46.

207. [Sphæria] Gossypii Sz.

S. simplex sparsa immersa globosa purpureo-incarnata submollis, ostiolo ad superficiem elongato gelatinam fundente.

Vix capsula Gossypii invenitur sine hac Sphæria. Minuta. Nascitur in capsulis Gossypii immaturis, primum profunde immersa, sed ostiolo etiam tum ad superficiem elongato, concolori, unde spargitur gelatina indurescens, quæ capsulis purpureo-rubro inquinat. Sphærulæ non observantur nisi capsula percissa; demum assurgunt et superficiem tum exsiccatam granulosam reddunt.

This species de Schweinitz placed in his Family VIII, Simplices, Sec. C, Brachystomæ, along with *S. putaminum* on peach pits and some others.

(2) Von SCHWEINITZ. Syn. N. Am. Fungi, p. 217. Mere mention as follows:

*1652. 507. S. gossypii, L. v. S., Syn. Car. 207, F. 212, non in Pennsylv.

(3) FRIES. Systema Mycologicum, Vol. II, 1823, p. 488.

412. *S. Gossypii.* Sparsa, submollis, perithcciis immersis globosis, purpureo-incarnatis, ostiolo ad superficiem elongato gelatinam fundente.

Schwein. (!) l. c. n. 207.

Minuta, dubiæ affinitatis. Nascitur in capsulis Gossypii immaturis, primum profunde immersa, sed ostiolo etiam tum ad superficiem elongato concolori, unde spargitur gelatina indurescens, quæ capsulas purpureo-inquinat. Perithecia non observantur nisi capsula perscissa, demum assurgunt et superficiem tum exsiccatam granulosam reddunt. In capsulis Gossypii copiose. (v. s.)

(4) SACCARDO. Sylloge Fungorum, Vol. II, p. 457. Saccardo not knowing what to do with this species refers it doubtfully to Hyponec-

51

tria as *H. gossypii*, quoting with a slight rearrangement of words the description of Fries.

(5) ELLIS. North American Pyrenomycetes, p. 71. Ellis uses Saccardo's name, gives an incomplete translation of the Latin description, and adds:

We have seen no specimens of this species, but have received from Prof. F. L. Scribner a *Fusarium* on capsules of cotton from South Carolina, which may be the conidial stage.

(6) ELLIS. Notes on Some Specimens of Pyrenomycetes in the Schweinitz Herbarium of the Academy. Reprint from Proceedings of the Academy of Natural Sciences of Philadelphia, February 21, 1893, p. 11.

Sphæria Gossypii, Schw. Syn. Car. 207. This is an obscure thing. The inner membrane of the cotton boll is wrinkled or roughened in drying so as to give the appearance of minute perithecia, but there is no fruit nor even any real perithecia.

(7) CURTIS. In Dr. Farlow's herbarium, at Cambridge, Mass., is a fragment of cotton capsule labeled "Sphæria gossypii Schw." This came from the herbarium of M. A. Curtis, who received it from de Schweinitz. On the pocket in the handwriting of Mr. Curtis is a penciled memorandum to the effect that this is not a fungus. Neither the writer nor Dr. Farlow, who examined the specimen with him, could find any Sphæriaceous or Nectriaceous fungus or any Fusarium spores on this specimen.

(8) Dr. KARL STARBÄCK, who kindly examined for me the Schweinitzian material of this species sent to Dr. Fries, and now preserved in the Fries collection at Upsala, writes that no fungus whatever is present and adds: "Species Schweinitzii Sph. gossypii est typicus observationis et auctoris et Friesii."

(9) My own examination in July, 1899, of material preserved in the de Schweinitz herbarium in Academy of Natural Sciences of Philadelphia, led to no different result. The specimens were examined both by reflected and transmitted light, with a hand lens and with the compound microscope.

The specimen in the collection proper (books of Sphærias) is labeled "Sphæria gossypii L. v. S. and Fr. Salem." This pocket contained nothing but some dust particles, insect detritus, and fragments, the largest of which was less than 1 mm. in area. These dust particles and fragments showed no perithecia or Fusarium spores. No fungus with necks either long or short was to be seen. Traces of an unknown, colorless, very delicate mycelium were observed, and one Macrosporium spore. This pocket also contained the skin of a museum pest.

In a separate package labeled "Fungi | Cryptograms | fr. Dr. Schweinitz | to Collins," I found, however, an uninjured, well-preserved specimen labeled in the handwriting of de Schweinitz "Sphæria Gossypii L. v. S. and Fr. Salem." This pocket contains two fragments of cotton capsule pericarp, each measuring about 1 by 1½ centimeters. The inner membrane is white, longitudinally wrinkled, and sound, or at least bears no Sphæriaceous bodies, Fusarium spores, or fungous threads. In places it bears tiny rusty specks, which are dead cells of the membrane containing some amorphous brown substance. The dark brown outer membrane is raised into numerous small papillæ, quite regularly arranged. These papillæ are barely visible to the naked eye and under a lens magnifying only two to three times might readily be mistaken for buried perithecia. These are undoubtedly what de Schweinitz saw and named *Sphæria gossypii*, but they do not contain any perithecia. Under a lens magnifying ten diameters they look more doubtful and when examined under the compound microscope (dry and crushed in water) they are observed clearly to be not of fungous origin, at least there are no Nectriaceous or Sphæriaceous bodies either on the surface or in the depths, and no Fusarium spores or hyphæ. Such papillæ are very common on the surface of cotton bolls when shriveling, as every one knows who has seen much of the plant, and very often at at least, they are not of fungous origin.

The presence of long necks which come to the surface and pour out gelatin and the statement by de Schweinitz that "scarcely a capsule of cotton is to be seen without this Sphæria" make it reasonably certain that de Schweinitz did not found his description on the infrequent little red perithecia which I have discovered. Finally, an examination of some of the scanty material of *Sphæria gossypii* which has been preserved in herbaria seems to indicate that the species was founded on the papillate appearance of the dry cotton capsule and on the fact that the surface of cotton capsules frequently exhibits a purple stain and a shining appearance. All the rest of the description (much more of it than I at first supposed) is pure inference—i. e., an easy method of accounting for the papillæ and the purplish glaze.

DESCRIPTION OF PLATE I.

1. Mature perithecia resting on fragment of hypocotyl of an old watermelon plant killed by the internal fungus. Monetta, S. C., September 9, 1895. This figure will answer equally well as an illustration of the perithecia occurring on cotton or cowpea.

2. Immature ascus, with two paraphyses and a few cells of the hypothecium, crushed out of a perithecium on watermelon. The smooth and still uncolored spores are surrounded by granular periplasm. Monetta, S. C., September 9, 1895.

3. Mature ascus taken from a perithecium which grew on mycelium derived from the ascospore shown in *6*. The periplasm has disappeared and the spores are now brown and have a thick wrinkled epispore.

4. A group of immature asci crushed out of a perithecium on cowpea.

5. Ripe ascospores of the watermelon fungus highly magnified.

6. Germinating ascospore, cowpea fungus. Agar-plate culture. James Island, S. C., August 22, 1895.

7. Conidia-bearing mycelium developed from the ascospore shown in *6*.

8. Three of the same conidia more highly magnified, one germinating. These conidia are identical in appearance with those produced inside of the vascular system of the still living stems. (See Plate II, *11*.)

9. Surface or dry-air conidia of the cowpea fungus. These were taken from the conidia beds shown in *11* and *12*.

10. Fragment from a cross section of the living cowpea stem, about 2 feet from the ground, showing a group of vessels infested by the fungus (*c*, cambium; *x*, xylem; *p*, pith), imbedded in paraffin, cut on the microtome, and stained for many hours in acid hæmatoxylin. This picture represents about one-thirtieth of the vascular ring, nearly all of which was occupied in the same manner. The cortical parenchyma and the bulky pith were still free from the fungus, as is always the case in this stage of the disease. Later the fungus pushes through to the surface, as shown in *11* and *12*. James Island, S. C., August, 1895.

11. Surface of dead stem of cowpea, showing rows of conidia beds. The vessels of such plants are always previously occupied by the internal fungus, as shown in *10*. The row-like arrangement of the conidia beds is due to the fact that the fungus comes to the surface along lines of least resistance—i. e., through parallel rows of parenchyma cells separating the strands of tough bast fibers (stereome). ×2. James Island, S. C., August, 1895.

12. A similar fragment of dead stem of the cowpea more highly magnified. ×10.

13. A tube of boiled rice overgrown by the mycelium of the watermelon fungus (*Fusarium niveum*). Culture No. 4, October 8, 1895. Painted October 28, 1895. This culture was bright blue on the start, 2 c. c. of violet litmus water and a few drops of a saturated solution of sodium carbonate (t. 25° C.) having been added to it. The same brilliant colors may be obtained, however, as already stated in the text, without use of litmus—i. e., by simply cultivating the fungus for a few weeks on rice boiled in distilled water—and this figure will answer equally well for such cultures. High up on the walls of the tube (above the rice) the fungus is white.

14. An incipient perithecium developing on mycelium produced by the ascospore shown in *6*.

With exception of *10*, *11*, and *12*, which were painted under my supervision by Mr. John L. Ridgway directly from the specimens, and *13*, which was painted in the same way by Miss D. G. Passmore, the figures are from my camera drawings. *1* was transferred to the plate and painted by Miss Passmore. The rest of the work was done by Mr. Ridgway. Where no measurements are given they are included in the plates or may be learned from the text or from other figures of the same sort in which they are given.

54

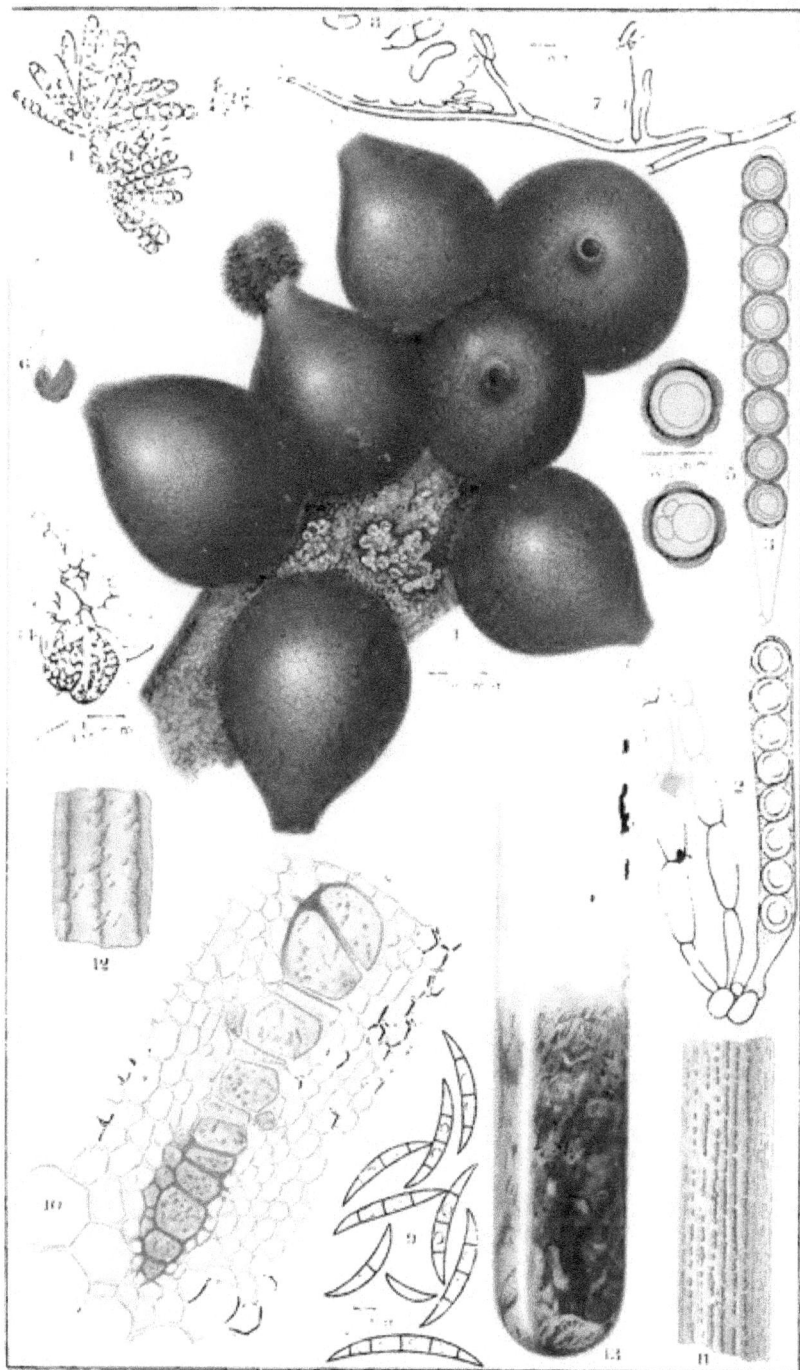

NEOCOSMOSPORA NOV. GEN.

ERWIN F. SMITH

DESCRIPTION OF PLATE II.

1. Closed stoma. Perithecium of the cowpea fungus. Diameter of stoma proper, 15 *µ*. James Island, S. C., August 29, 1895.

2. Open stoma of the cowpea fungus, showing periphyses lining the inner wall of the throat. Diameter of the stoma, 30 *µ*. James Island, S. C., August 29, 1895.

3. Another view of the periphyses, a portion of the inner wall of the neck of the perithecium of the cowpea fungus crushed out and examined in water. James Island, S. C., August 29, 1895.

4. Optical section through perithecium from a dead watermelon stem, showing cavity full of loose ascospores. Some of the asci remained, but the walls of most had dissolved. This figure will answer equally well for the cotton or cowpea fungus in each of which the same phenomenon was observed. Size, 320 by 280 *µ*. Slightly diagrammatic. Monetta, S. C., September 9, 1895.

5. Peridial cells in the middle or ventral portion of a perithecium of the cowpea fungus. The same are shown less distinctly and somewhat diagrammatically in Plate I, *1*, and Plate V, *1*. James Island, S. C., August 27, 1895.

6. Conidial tufts on the surface of a watermelon stem killed by the internal fungus. Monetta, S. C., July 16, 1894.

7. Top of a similar tuft (melon fungus), showing attached and loose spores in various stages of growth and septation. From a plant destroyed by soil infection. Washington, D. C., September 27, 1894.

8. Parenchyma cells from the living hypocotyl of a young watermelon, third day of the wilt. Parenchyma partially occupied by conidia-bearing mycelium. No fungus on the surface. In this cell there were 26 conidia. The section was jarred repeatedly and somewhat roughly, and finally turned over without disturbing the fungus. Undisturbed nonparasitized parenchyma cells were above and below it, and the conidia were plainly inside of the cell here figured, which was from tissue near a fungus-infested bundle. Hothouse experiment, Washington, D. C., April 10, 1895.

9. An external, lunulate, 3-septate conidium of the watermelon fungus after seventeen hours in acid cucumber agar. Twenty-four hours later numerous elliptical conidia, like those shown in Plate I, *7*, were abstracted.

10. Hypha end, showing abstriction of the internal conidia from mycelium of 9, forty-eight hours after the latter was drawn.

11. Internal conidia of cowpea fungus. These spores were taken from the vessels of cowpea stems on James Island, S. C., August 7, 1895. Twenty-nine internal conidia were measured that day, the size varying as follows: Length, 4.5 to 18 *µ*; breadth 2.3 to 4 *µ*.

Figures transferred to the plate by Mr. Williams Welch, from camera drawings by the author.

56

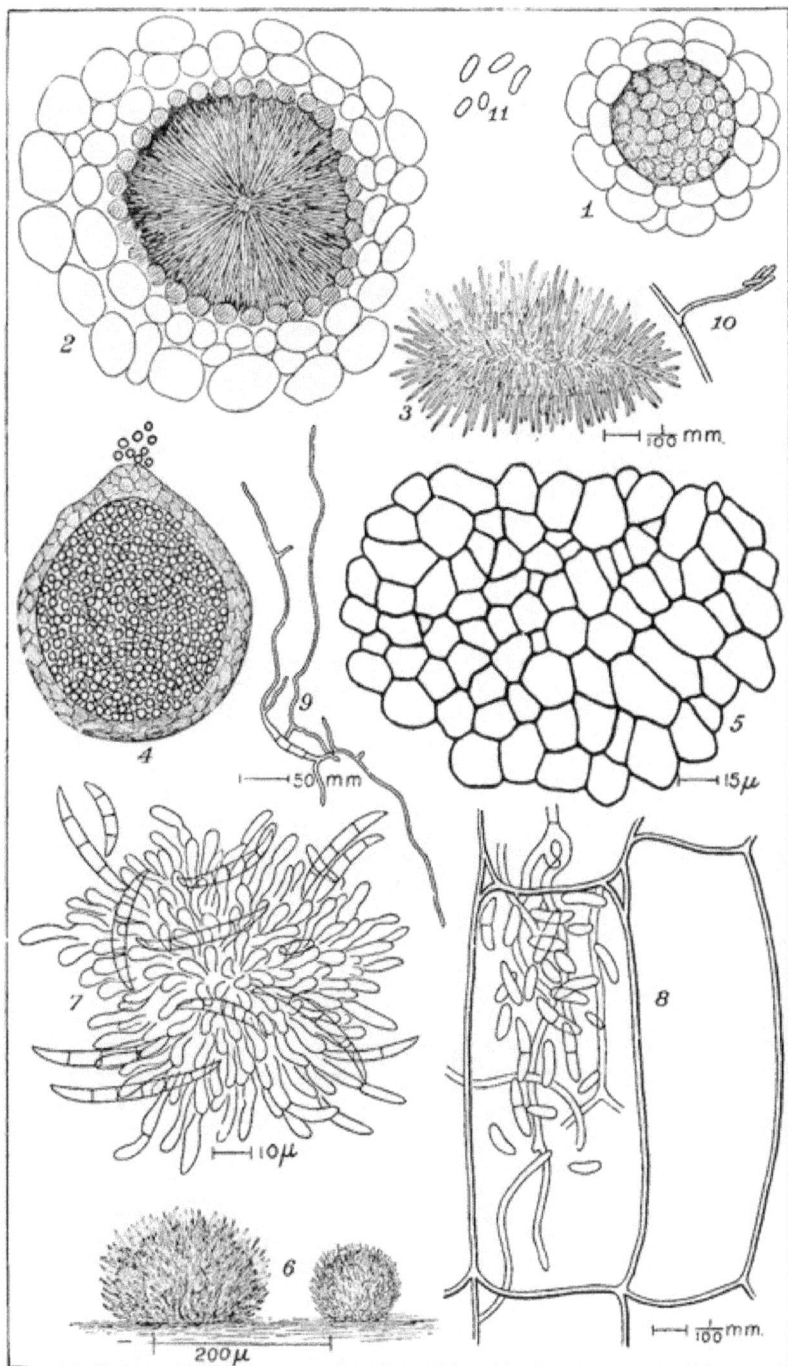

FUNGUS OF THE MELON AND COWPEA WILT.

1. Microconidium from the interior of a melon stem, germinating after 7 hours in agar.

2. The same conidium, after 24 hours in agar, showing numerous branches and the abstriction of new conidia. Under favorable conditions conidia are formed very rapidly. Three hours after this drawing was made this mycelium had given rise to 40 free conidia and 30 more were in various stages of growth.

3. Formation of microconidia of the watermelon fungus in an agar plate culture. This hypha end was under continuous observation for two hours and eighteen minutes, during which time two spores were developed and abstricted and another begun. Room temperature, 27° C. The formation of conidia was carefully followed in a number of other cases. Under the most favorable conditions of temperature and food supply only forty-five minutes intervened between the pushing off of one conidium and the formation and abscission of another. Usually, however, fifty-five to sixty minutes was required.

4. Mycelium and conidia of the melon fungus from a young agar culture. This was one piece of mycelium, broken in the drawing at the place marked x for convenience of representation on the plate. Mycelium and conidia vacuolate, all formed in the agar, only sterile hyphæ ends projecting into the air in this early stage of growth. Two spores germinating. This figure well illustrates the way the hypha end throws off a few conidia, passes into a vegetative condition, and elongates for a time, with formation of septa, and then once more ceases to elongate and becomes sporiferous. Monetta, S. C., July 4, 1894.

5. Microconidia of the melon fungus from an agar plate, showing the variability in size. Two spores germinating.

6. Conidia and torulose mycelium from a culture of the internal watermelon fungus (*Fusarium niveum*) 27 days old.

7. Fragment of mycelium from the same culture as 6.

8. Fragment of mycelium from a culture of the internal cotton fungus (*Fusarium vasinfectum*) 25 days old. For comparison with 7.

9. Mycelium and microconidia of the cotton fungus (*Fusarium vasinfectum*), cultivated from the interior of a diseased cotton stem received from western Georgia. From a pure white culture 25 days old. For comparison with 2 and 4.

10. Conidia of the cotton fungus from a culture 25 days old, derived from the internal or microconidia. For comparison with 11.

11. Macroconidia and microconidia of the watermelon fungus from a pure culture 5 months old, on sterilized horse dung. The mycelium which bore these spores was derived from a spore of the size and shape of the largest here shown.

12. Chlamydospores of the melon fungus. Several germinating. From a pure culture 5 months old, on horse dung. This culture was derived from a lunulate, several-septate, external conidium. In mass these chlamydospores were brick red, and their contents was considerably denser than has been indicated by the engraver. At 11 a. m., when the examinations began (in water), there were no germinations; at 1.30 p. m. there were many. In the same tube with these chlamydospores were conidia of all the sizes shown in 11, the small spores being much more numerous than the large ones.

Figures drawn by the author and engraved on wood by L. S. Williams.

58

PLATE III.

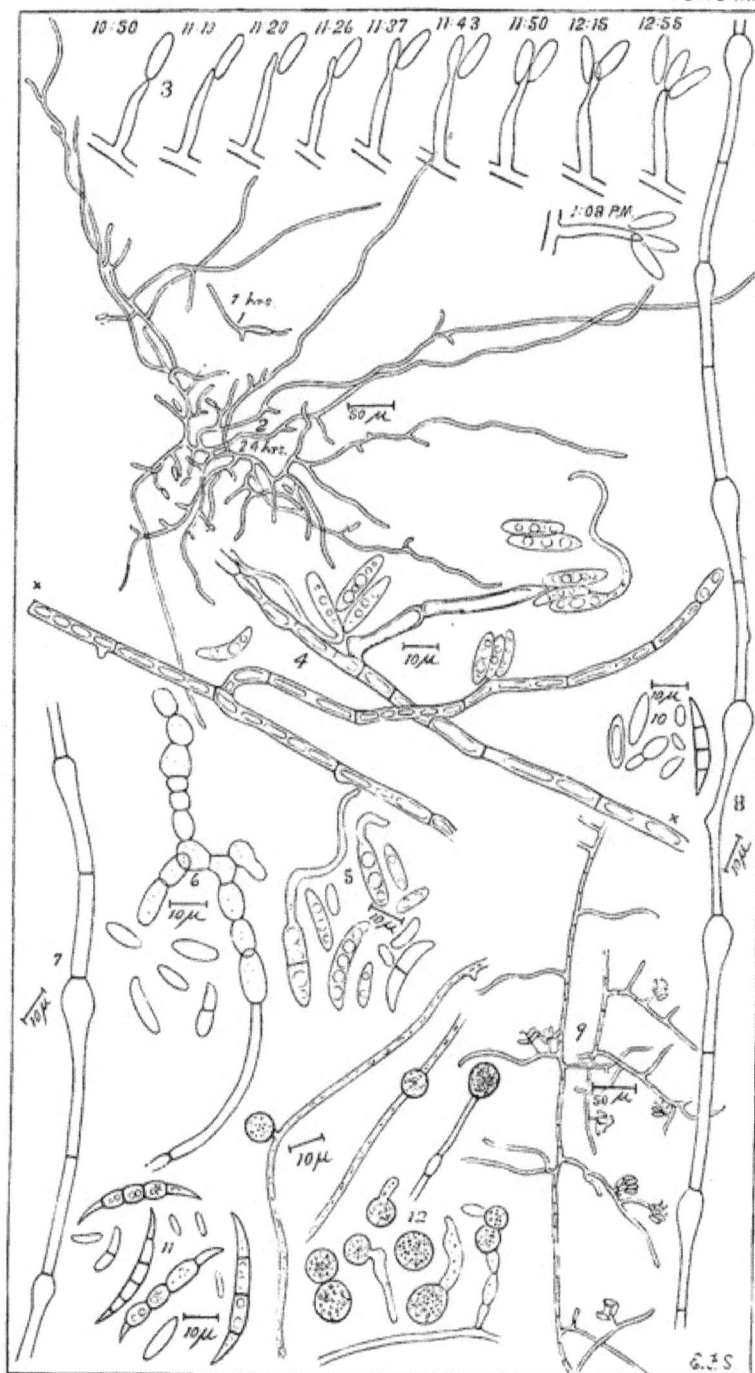

FUNGUS OF THE MELON AND COTTON WILT.

(From culture media.)

Cross section of a mature watermelon stem, showing how the vessels are plugged by the fungus. In this stage of the disease the foliage has suddenly wilted (as shown in Pl. VII, 1), but is not yet shriveled; the stem is green and turgid and its parenchyma is not yet invaded. From the surface inward the tissues are as follows: (1) Epidermis, (2) collenchyma, (3) cortical parenchyma, (4) bast fibers (stereome), (5) medullary system, (6) ten bicollateral bundles in two rows, (7) pith. The structure of the bundles is as follows: (1) outer phloem, (2) cambium, (3) xylem, consisting of pitted and reticulated vessels held together by wood parenchyma, (4) spiral vessels lying in non-lignified living parenchyma (the primary vessel parenchyma of Strasburger), (5) pseudocambial layer, (6) inner phloem, composed like the outer phloem of sieve tubes and companion cells. On the outer side of the outer phloem may be seen the collapsed remnants of the primary sieve tubes. The middle portion of the stem is occupied by fissures. Section embedded in paraffin, cut on the microtome, and stained in hæmatoxylin. Reduced one-third from a pen drawing made directly from the section by Mr. W. Scholl. Diameter of the stem, 4 millimeters.

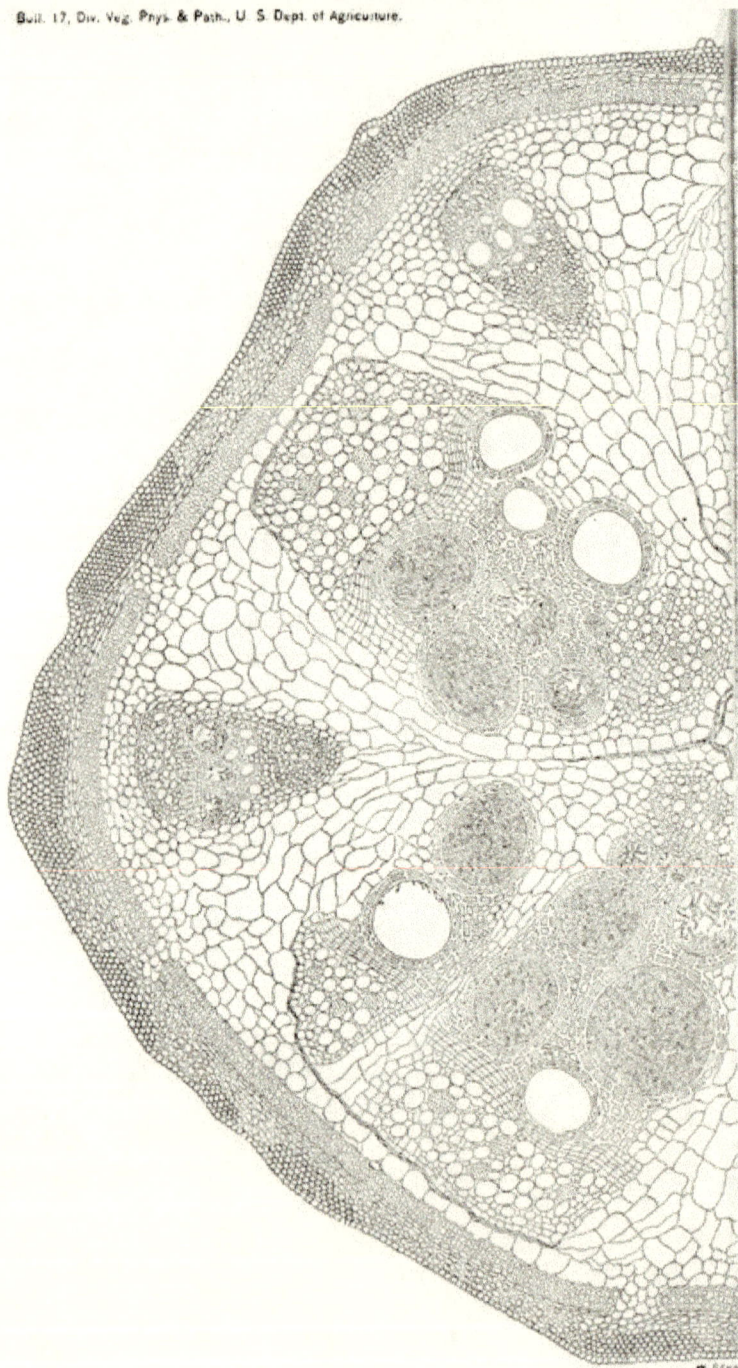

THE WATE
Cross section of a stem, showing how

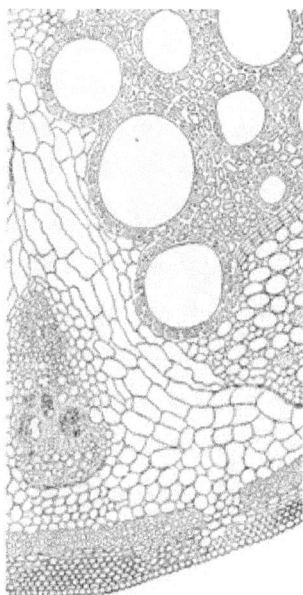

ON WILT.

vessels are plugged by the fungus.

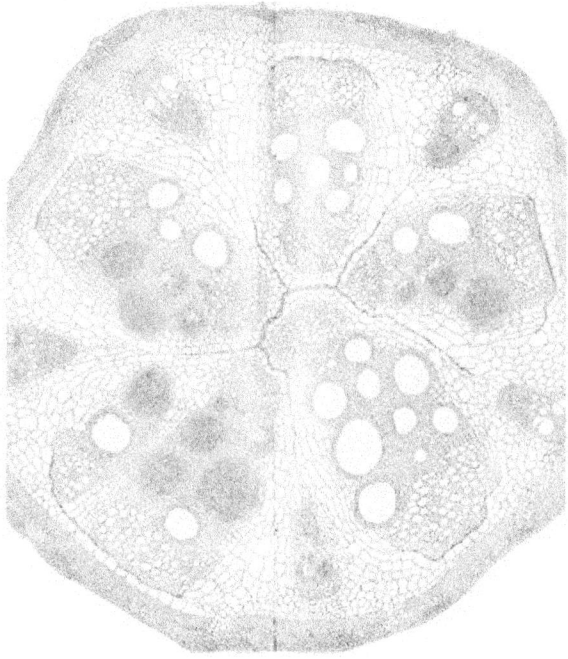

1. Perithecium and ascospores from upland cotton. Salters Depot, Williamsburg Co., S. C., Oct. 8, 1895. Size 352 by 272 μ. For comparison with Pl. I, *1*, which, however, is less highly magnified. The ascospores of this perithecium were 10 by 10 μ with a wrinkled exospore. Others on the same specimens were 9 by 9 μ. More rarely they were 9 by 10 μ or 9 by 11 μ.

2. Ripe ascus and paraphysis from the same lot of specimens as *1*. A paraphysis from a perithecium on cowpea was 18 μ broad (cell next to the end cell).

3. Macroconidia from conidia beds on the surface of killed stems of sea-island cotton. James Island, S. C., June 29, 1895. All transitions between *a* and *b* were observed.

4. Internal or microconidia from a diseased okra plant. James Island, S. C., July 25, 1895. Size 8 to 16 by 2.5 to 3.5 μ. This okra had been planted in place of cotton which wilted and died early in the growing season. Many plants were affected. The stems were 2 feet tall and about 1 inch in diameter at the base. The wood was much browned the whole length of the stem.

5. Macroconidia from external conidia beds on dead okra plants. James Island, S. C., July 25, 1895. The mature spores measured 27 to 42 by 3 to 5 μ. Occasionally one was 5-septate. The vessels and parenchyma of these plants contained a great amount of mycelium bearing such spores as are shown in *4*.

6. Highly magnified cross section of a pitted vessel which is beginning to be occupied by the melon fungus. Stem of watermelon. Monetta, S. C., June 26, 1891. Camera drawing from a fresh section examined in water. A great many of the vessels were plugged solid by the fungus, as if stuffed with cotton. For the location of these fungus-infested vessels with reference to other parts of the stem see Plate IV.

Figures transferred to the plate by Mr. Williams Welch from camera drawings by the author.

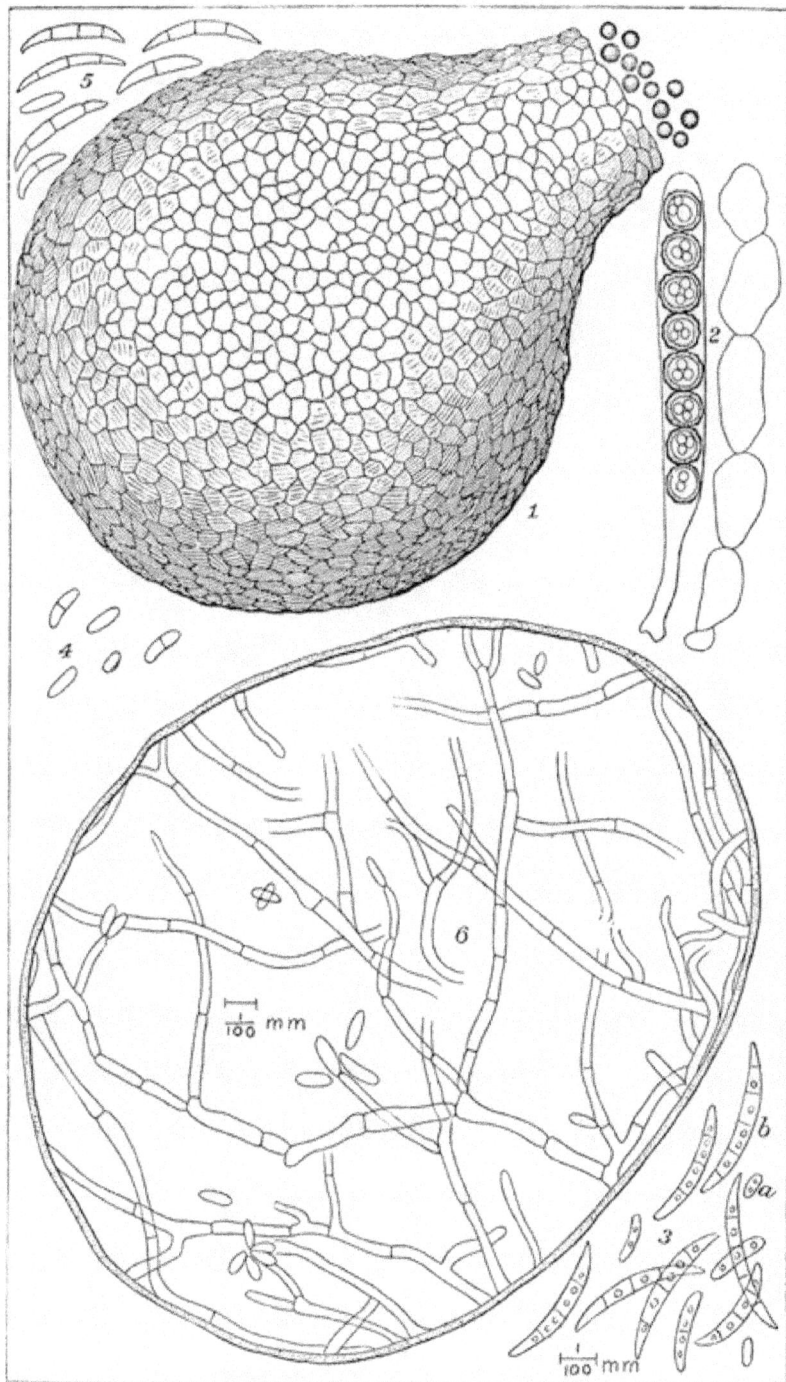

FUNGUS OF THE COTTON, OKRA, AND MELON WILT.

DESCRIPTION OF PLATE VI.

Healthy melon vines. Monetta, S. C. June 28, 1894. For comparison with the photographs shown on the following plate, which were made at the same time and from the same field. The vine in the foreground had about 200 leaves and a melon three-fourths grown.

64

HEALTHY WATERMELON VINES.

Marietta, S. C., June, 1891.

DESCRIPTION OF PLATE VII.

Vines attacked by the melon fungus. Monetta, S. C. Photographs a. m. of June 28, 1891.

1. On the afternoon of June 26, when last examined, this vine was to all appearances as healthy as that shown on Plate VI.

2. A vine which has been wilted for several days. Healthy vines in the background.

66

WILTED MELON VINES FROM THE SAME FIELD AS PLATE VI.

DESCRIPTION OF PLATE VIII.

Melon wilt. The result of soil inoculations. Hothouse experiment. Washington, D. C. Photographed April 17, 1895. Soil inoculated in 1894 with internal melon fungus brought from South Carolina.

1. Two healthy and 3 diseased plants; *2.* One healthy and 3 diseased plants.

In this stage of growth the first symptom is a drooping of the cotyledons; this is followed by flaccidity of the plumule and a bowing over of the hypocotyl. The parts above ground and the roots also are sound externally—i. e. they are not wounded, rotted, softened, shriveled, or browned in any way. In this early stage of the disease there is little or no fungus in the hypocotyl, but the vessels of the taproot are plugged.[1] After photographing, water was withheld from these plants, and as they were in a dry room they soon died, the healthy ones included. On April 24 each of the 6 plants which were wilted when photographed bore the conidia beds of the external Fusarium, and a further examination showed the bundles and parenchymatic tissues of these plants to be full of the internal mycelium and microconidia. On the contrary, the 3 plants which were healthy when photographed contained no internal fungus, and there were no Fusarium beds on the surface, although the plants were under a bell jar in moist air for a day or two prior to the examination.

---

[1] June 20, 1894, about 1,500 hills of watermelons were planted by the writer on a sandy field in South Carolina, which had been infected from stable manure, and on which most of the melon vines had been destroyed by this fungus in May and June of that year. The disease began soon after the plants came up, and in 6 weeks nearly all of the young melons had wilted, altogether perhaps eight or ten thousand plants. The cotyledons first became flabby and drooped, the first true leaf then wilted, the hypocotyl lost turgor and bowed over to the ground, and the plants finally shriveled—i. e., the symptoms were precisely the same as those subsequently obtained in Washington with pure cultures of the melon fungus. A hundred or more of these wilting plants were examined microscopically, and in each one the fungus was found in the vessels of the hypocotyl or taproot or both in quantity sufficient to account for the disease and commonly nothing else was present. In many plants which were not pulled until the second or third day of the wilt, the fungus was found pushing out into the parenchyma cells and fruiting therein, as shown on Pl. II, *8.* July 14 the writer removed 14 healthy-looking young melon plants from as many different hills in this field and examined them microscopically for the presence of the fungus. Vines had recently wilted in each of these hills. The big seedlings were growing rapidly and appeared to be perfectly healthy above ground and below. In 12 of these plants no fungus could be found. In 1 there was an abundance of the fungus in the big central duct of the taproot and in some of the smaller vessels, but none could be found in the hypocotyl. In the other, there was also no fungus in the hypocotyl save doubtfully a thread or two in one vessel, but there was plenty of it in the big ducts of the taproot about 1 centimeter below the crown. No hyphæ threads were observed in the parenchyma cells of the roots, which were white and appeared to be entirely sound. These two plants would have wilted in a day or two, and the history of the experiment shows that the other 12 would subsequently have contracted the disease.

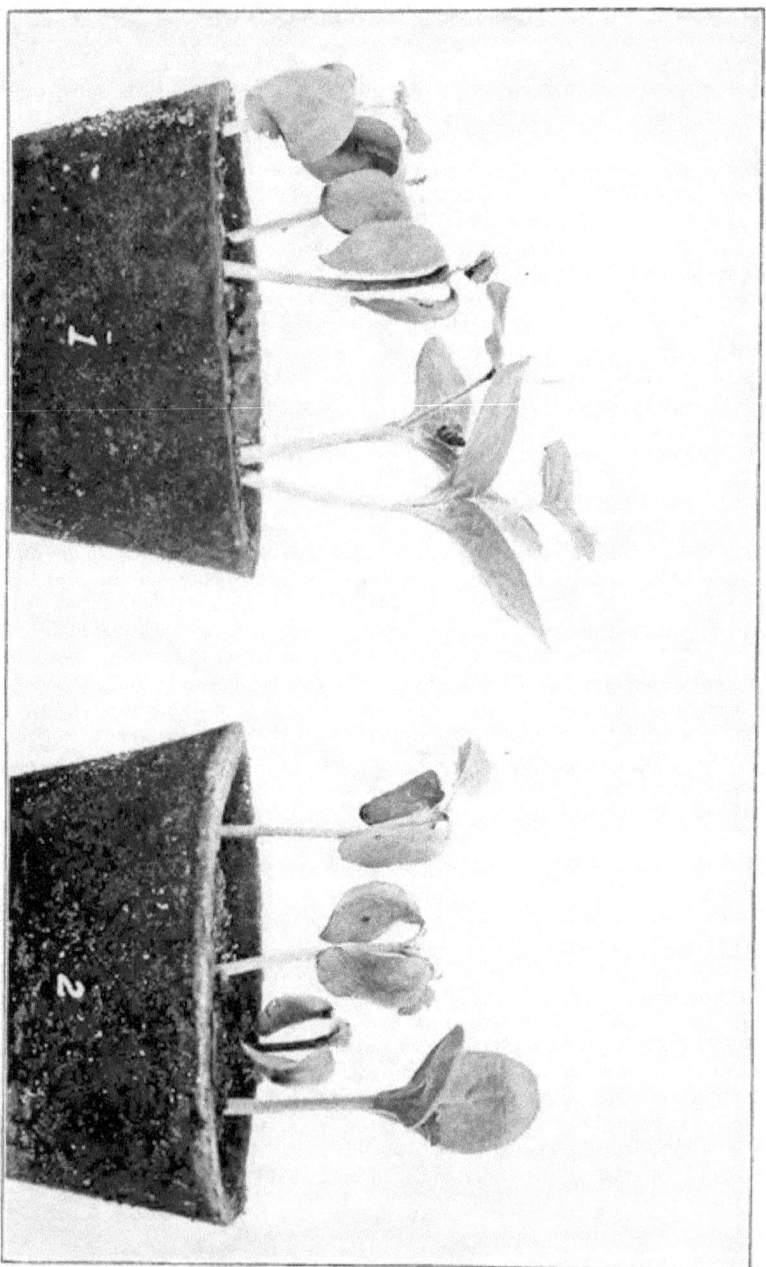

THE MELON WILT IN SEEDLINGS.

DESCRIPTION OF PLATE IX.

Watermelon wilt. Hothouse experiment. Washington, D. C. Spring of 1895.

1. Pots of the same series as Pl. VIII, but photographed some weeks later. All of the plants were killed by the fungus except the three mentioned above. Seeds planted March 12; photograph made some time in May.

2. Control plants growing in uninoculated (healthy) soil. None of these plants contracted the disease, but toward the close of the experiment all were dwarfed by leaf aphides.

70

THE MELON WILT.

Soil inoculated with mycelium from an *infected* condition. (Fungus now fruiting on the surface of the dead stems.)

Melon wilt. Result of soil inoculations *with the external fungus.* Hothouse experiment. Washington, D. C. Photographed June 10, 1895. The fungus used for infection was derived from a single, large, lunulate, several-septate spore, taken from the surface of a plant killed by the internal fungus. When photographed there were no conidia beds on the surface of these wilted plants, but in each case the internal fungus was found plugging the vessels of the taproot; the internal conidia were elliptical or bluntly pointed, straight or slightly curved, non-septate, and 8 to 20 by 3.5 to 4 μ. On June 8 the plant at the right was as healthy looking as its companion, but the plant in the middle pot and at the extreme left were already wilted. The three healthy vines afterwards contracted the disease.

72